SpringerBriefs in Food, Health, and Nutrition

Editor-in-Chief
Richard W. Hartel
University of Wisconsin—Madison, USA

Associate Editors
J. Peter Clark, *Consultant to the Process Industries, USA*
John W. Finley, *Louisiana State University, USA*
David Rodriguez-Lazaro, *ITACyL, Spain*
Yrjö Roos, *University College Cork, Ireland*
David Topping, *CSIRO, Australia*

Springer Briefs in Food, Health, and Nutrition present concise summaries of cutting edge research and practical applications across a wide range of topics related to the field of food science, including its impact and relationship to health and nutrition. Subjects include:

- Food Chemistry, including analytical methods; ingredient functionality; physic-chemical aspects; thermodynamics
- Food Microbiology, including food safety; fermentation; foodborne pathogens; detection methods
- Food Process Engineering, including unit operations; mass transfer; heating, chilling and freezing; thermal and non-thermal processing, new technologies
- Food Physics, including material science; rheology, chewing/mastication
- Food Policy
- And applications to:
  - Sensory Science
  - Packaging
  - Food Quality
  - Product Development

We are especially interested in how these areas impact or are related to health and nutrition.

Featuring compact volumes of 50 to 125 pages, the series covers a range of content from professional to academic. Typical topics might include:

- A timely report of state-of-the art analytical techniques
- A bridge between new research results, as published in journal articles, and a contextual literature review
- A snapshot of a hot or emerging topic
- An in-depth case study
- A presentation of core concepts that students must understand in order to make independent contributions

For further volumes:
http://www.springer.com/series/10203

Ioannis Arvanitoyannis • Achilleas Bouletis
Dimitrios Ntionias

Application of Modified Atmosphere Packaging on Quality of Selected Vegetables

Springer

Ioannis Arvanitoyannis
Department of Ichthyology and Aquatic
  Environment
School of Agricultural Sciences
University of Thessaly
Hellas, Greece

Achilleas Bouletis
Department of Ichthyology and Aquatic
  Environment
School of Agricultural Sciences
University of Thessaly
Hellas, Greece

Dimitrios Ntionias
Department of Ichthyology and Aquatic
  Environment
School of Agricultural Sciences
University of Thessaly
Hellas, Greece

ISSN 2197-571X   ISSN 2197-5728 (electronic)
ISBN 978-3-319-10231-3   ISBN 978-3-319-10232-0 (eBook)
DOI 10.1007/978-3-319-10232-0
Springer Cham Heidelberg New York Dordrecht London

Library of Congress Control Number: 2014949239

Printed on acid-free paper

Springer is part of Springer Science+Business Media (www.springer.com)

Introduction

A large demand for fresh, healthy, and convenient food has led to a tremendous growth of ready-to-use (RTU) vegetable industry, as fresh cut fruits and vegetables are defined products that have been trimmed and/or peeled and/or cut into 100 % usable product that is offered to the customer and retains high nutrition value, convenience, and flavor (Rico et al. 2007).

Respiration of vegetables can be defined as the activity of degradation of composite organic compounds such as sugars, organic acids, amino acids, and fatty acids for the final result of energy production through oxidation. The enzymic reactions are called respiration (Ooraikul and Stiles 1990). Ethylene has a leading role on the metabolic activities of the plant, increasing respiration rate and accelerate ripening and senescence (Nguyen-the and Carlin 1994; Lin and Zhao 2007). Transpiration is the loss of humidity from plant tissues. Water loss can lead to a serious quality loss. Enzymic browning can be treated as one of the main causes for quality loss in fruits and vegetables. In lettuce, altered phenol metabolism plays an important role in tissue browning (Saltveit 2000).

Contents

Abbreviations

1-MCP	1-Methylcyclopropene
AM	Automatic misting
AP	Aerobic population
BA	Benzylaminopurine
BOPP	Bioriented polypropylene
CA	Controlled atmosphere
CAS	Controlled atmosphere storage
CAT	Catalase
CFU	Colony forming units
Chl	Chlorophyll
DHAA	Dehydrated ascorbic acid
DPPH	2,2-Diphenyl-1-picrylhydrazyl
DW	Dry weight
EAP	Ethylene absorber
EMA	Equilibrium modified atmosphere
ES	Ethylene scavenger
FW	Fresh weight
GAC	Granular activated carbon
GAD	Glutamate decarboxylase
HAV	*Hepatitis A* virus
HOA	High oxygen treatment
HWT	Hot water treatment
IPRH	In package relative humidity
KMS	Potassium metasulphite
LAB	Lactic acid bacteria
LDPE	Low density polyethylene
LLDPE	Linear low density polyethylene
MA	Modified atmosphere
MAP	Modified atmosphere packaging
Ma-P	Macro-perforated
MDA	Malondialdehyde

Mi-P	Micro-perforated
MVP	Moderate vacuum packaging
No-P	Non-perforated
OPP	Oriented polypropylene
OTR	Oxygen transmission rate
OVQ	Overall visual quality
PAL	Phenylalanine ammonia lyase
PE	Polyethylene
PET	Polyethylene terephthalate
PLA	Polylactic acid
PMAP	Passive modified atmosphere packaging
POD	Peroxidase
PP	Polypropylene
PPB	Parts per billion
PPM	Parts per million
PPO	Polyphenol oxidase
PVC	Polyvinyl chloride
RTE	Ready to eat
SH	Sodium hypochlorite
SOD	Superoxide dismutase
SS	Sodium sulphite
SSC	Soluble solids content
TA	Titratable acidity
TAA	Total ascorbic acid
Tchl	Total chlorophyll
TF	Temperature fluctuation
TSS	Total soluble solids
UV	Ultra violet
VC	Vacuum cooling
VP	Vent packaging

Chapter 1
Application of Modified Atmosphere Packaging on Quality of Selected Vegetables: A Review

1.1 Chemical and Microbiological Parameters Affecting Shelf Life

The most common persistent organic pollutants (POPs) found were Dichlorodiphenyl trichloroethane (DDT) and its metabolites (found in 21 % of samples tested in 1998 and 22 % in 1999), and dieldrin (found in 10 % of samples tested in 1998 and 12 % in 1999) while residues of five and more chemicals could be found in a single food item (Schafer and Kegley 2002). Lettuce was shown to have the highest concentration of fungicides while Swiss chard appeared to have the highest levels of insecticides. Washing, blanching, and peeling of vegetables led to a 50–100 % reduction in some cases of pesticide levels (Chavarri et al. 2005). Cadmium, copper, lead, chromium, and mercury pose an important threat and can be detected in high concentrations in vegetables (Islam et al. 2007).

The use of any microbial preventive technique is excluded in the case of minimally processed vegetables because of the serious impact on their quality (Zagory 1999). Growth in open spaces and postharvest handling create an elaborate spoilage microflora on vegetables due to contact with various types of microorganisms (Villanueva et al. 2005). The type of the vegetable can determine both the type of spoilage and subsequent quality deterioration (Jacxsens et al. 2003). *Pseudomonas*, *Enterobacter* and *Erwinia* are the predominant species, while Gram$^{(-)}$ amount to 80–90 % of bacteria detected (Francis et al. 1999).

Asparagus, broccoli, cauliflower, carrot, celery, cherry tomatoes, courgette, cucumber, lettuce, mushroom, pepper, turnip, and watercress are some of the vegetables in which *Aeromonas* spp. was isolated (Merino et al. 1995). Mushrooms from retail markets in United States and various produce items sampled from farmers' markets in Canada were contaminated with *Campylobacter* (Harris et al. 2003). The assembly of ready-to-eat (RTE) meals containing beef or other potential carriers of *E. coli 0157:H7* combined with raw salad vegetables may lead to their contamination

© The Author(s) 2014
I. Arvanitoyannis et al., *Application of Modified Atmosphere Packaging on Quality of Selected Vegetables*, SpringerBriefs in Food, Health, and Nutrition, DOI 10.1007/978-3-319-10232-0_1

(Abdul-Raouf et al. 1993). A number of surveys stated the presence of *Listeria monocytogenes* on cucumber, peppers, potato, radish, leafy vegetables, bean sprouts, broccoli, tomato, and cabbage at sailing points (Heaton and Jones 2008).

1.2 Modified Atmosphere Packaging (MAP)

MAP is a technique for extending the shelf life of fresh and minimally processed food. During the atmosphere modification procedure the air surrounding the product is replaced by the desired composition. MAP is applied in a wide variety of products and the mixture of gases used are selected and affected by many factors such as storage temperature, type of the product and packaging materials (Sandhya 2010).

In MAP/Controlled atmosphere (CA) oxygen, carbon dioxide and nitrogen are the most common gases used. Nitrous and nitric oxides, sulfur dioxide, ethylene, chlorine, as well as ozone and propylene oxide have been used in many experiments to investigate their effect on the postharvest quality of vegetables. The afore-mentioned gases did not find commercial use due to high cost and safety and regulatory considerations. The compositions of gases used in MA are, inert blanketing using N_2, semi-reactive blanketing using CO_2/N_2 or $O_2/CO_2/N_2$, or fully reactive blanketing using CO_2 or CO_2/O_2 (FDA/CFSAN 2001).

The use of MAP can result in reduction of respiratory activity, retardation of softening and ripening and restraint of pathogens and reduced incidence of various physiological disorders (Caleb et al. 2012).

In all food preservation techniques there are critical parameters that play an important role in shelf life prolongation, functionality and effectiveness of the method. The effect of storage temperature, gas composition, the nature of the products and the wrapping film is important to be defined and evaluate any interactions in order to achieve the optimal results of modified atmosphere packaging (Arvanitoyannis and Bouletis 2012).

1.3 Temperature

Temperature is an extremely important factor during packaging design due to its effect on the physiology of the product. One of the main responses to stress is protein dysfunction, disrupting cellular homeostasis, known as heat shock proteins. The production of these proteins is triggered by conditions like oxidative stress, low temperatures and fruit ripening. So exposure of sensitive products to low storage temperatures may have beneficial effect on shelf life and preserve quality but there are some limitations. If these limits are exceeded may lead to chilling injury (Aghdama et al. 2013).

Preserving the ideal storage temperature in all sections of the postharvest handling chains is vital for the quality and the shelf life of the product. When truck or sea transportation is involved, temperature limits can be attained but the extended transport time may be prohibitive for short-life products to reach all the available markets. On the other hand, air transport offers limited transportation time but a significant transgression of temperature limitations is involved both during flight and ground operations (Brecht et al. 2003).

1.4 Gas Composition

MAP conditions are usually used to prevent microbial growth and ensuing spoilage. CO_2 is used for this purpose either alone (in 100 % concentrations) or combined with O_2 and N_2 and plays an important role in shelf life extension of the product by preventing growth of aerobic microflora (Puligundla et al. 2012).

Oxygen is necessary for several processes that lead to quality degradation of the product such as fat oxidation, browning and pigment oxidation. Also an atmosphere rich in O_2 favors the growth of all aerobic bacteria and fungi. So, to avoid any of these situations most of atmosphere modifications contain low concentrations or absence of oxygen (Sandhya 2010).

N_2 is an inert gas used for filling the additional space in the package to prevent collapse. It has low solubility in water and lipid (Phillips 1996).

CO is a gas not very soluble in water but with increased solubility in organic solvents. Its use was permitted in the U.S. in packed lettuce due to its ability to browning prevention. The very limited applications in the food industry are a result of both its toxicity and its explosive nature (Sandhya 2010).

1.5 Nature of the Product

The process in which organic materials (carbohydrates, proteins and fats) are degraded into simple end products with simultaneous energy emission is called respiration. Through this process, CO_2 is a final byproduct and O_2 is consumed. Respiration continues even after harvest and during storage, so it is crucial that each type of vegetable interacts with the surrounding atmosphere and if the conditions are not satisfactory, physiological disorder occurs (Workneh and Osthoff 2010).

Ethylene (C_2H_4) is a plant hormone that significantly affects the physiological process of the plants. Storage temperature, injuries, disease occurrence and water absence induce ethylene formation. Rapid ripening and increased respiration leading to shortened shelf life are directly affected by ethylene exposure. Specific packaging conditions can lead to reduced ethylene formation and consequently to prolonged shelf life (Irtwange 2006).

Transpiration is another process that significantly affects post-harvest quality of the product. When the produce is harvested from the growing plant, the continuous water supply is interrupted and the only source for transpiration is the internal water content. So increased post-harvest transpiration can result in weight loss and shriveling, which may cause rejection of the product on the markets. Surface-to-volume ratio, injuries, morphological and anatomical characteristics, maturity stage and in package conditions such as temperature, relative humidity (RH), air movement and atmospheric pressure are some of the factors that can affect transpiration rate of fresh produce (Caleb et al. 2013).

1.6 Packaging Film

Polymeric compounds are the main materials for flexible package structures used for MAP but they also can be used on a rigid or semi-rigid packaging solution such as a lidding on a tray. Low-density polyethylene (LDPE), linear low-density polyethylene (LLDPE), high-density polyethylene (HDPE), polypropylene (PP), polyvinyl chloride (PVC), polyester, i.e. polyethylene terephthalate (PET), polyvinylidene chloride (PVDC), polyamide (Nylon) are some of the plastic films used in MAP (Mangaraj et al. 2009). Characteristics of the film that can affect atmosphere modification and must be carefully selected are film permeability to O_2, CO_2 and water vapor, film thickness, package surface area and free volume inside the package (Caleb et al. 2013).

The selection of the optimal packaging material for each vegetable depends on several factors:

1. The form of package (rigid tray or flexible pouch)
2. O_2, CO_2 and water vapor transmission rates.
3. Physical characteristics of film (clarity, durability, stretch capability, etc.)
4. Heat sealing ability, anti-fogging properties
5. Sealing properties
6. Resistance to chemical factors
7. Absence of toxicity and interaction with the product
8. It is a clear need for a printable, cheap and commercially suitable film (Mangaraj et al. 2009).

Edible coatings are thin layers of edible material that provide either protection or act as a barrier for moisture, oxygen and solute movement for the food and were first recorded in the early twelfth century (wax coating). The main action of edible coatings is the accumulation of CO_2 produced through respiration within the product that eventually leads to partial anaerobic conditions. So the product consumes less oxygen and ethylene production is restrained. Polysaccharides, proteins, lipids, and composites are some of the materials used for edible coatings. (Dhall 2013).

1.7 MAP Application on Vegetables

1.7.1 Roots

1.7.1.1 Lotus

Lotus roots were treated with chitosan coating and subsequently stored under MAP at 4 °C for 10 days. L^* values of treated samples differed significantly from the control (68.8 and 48.9 for treated and control samples, respectively after 8 days of storage). MDA (malondialdehyde) content (8.2 nmol/g after 8 days of storage) and PPO (polyphenol oxidase) activity (520.7 U/min mL at the end of storage) were lower compared to control (Xing et al. 2010).

1.7.1.2 Carrots

Growth potential of inoculated pathogens such as *Escherichia coli O157:H7*, was tested on carrots stored under PMAP (two OPP films were used with OTR of 3,500 and 1,100 cm^3 m^{-2} day^{-1} atm.$^{-1}$ for OPP 1 and OPP 2 respectively) at 5 and 25 °C for 8 days. *E. coli O157:H7* reached 7, 8 and 8.4 log CFU/g on samples stored in OPP 1, OPP 2 and control bags, respectively after 3 days of storage at 25 °C. At 5 °C, pathogens strains survived but did not increase their populations (Abadias et al. 2012).

Alasalvar et al. (2005) focused on shredded orange and purple carrots under MAP (95 % O_2/5 % CO_2, 5 % O_2/5 % CO_2) at 5 ± 2 °C. A major decrease in antioxidant activity (from 220 to 180 μmol TE/g) under specific conditions (95 % O_2 + 5 % CO_2) was reported for purple carrots. A significant reduction of the accumulation of total phenols resulted in better sensory quality and extended shelf-life for purple carrots (for at least 2–3 more days).

Amanatidou et al. (2000) studied the impact of high O_2 and high CO_2 MA (90 % O_2/10 % CO_2, 80 % O_2/20 % CO_2, 50 % O_2/30 % CO_2, 70 % O_2/30 % CO_2, 1 % O_2/10 % CO_2) in the presence of citric acid (0.1 or 0.5 %), hydrogen peroxide, chlorine, $CaCl_2$ and an alginate edible coating on the preservation of carrots. The original appearance of the carrots for 8 days at 8 °C was maintained by applying citric acid 0.5 % (w/v). Firmness of carrots and initial pH were not affected by the chlorine treatment. Satisfactory quality was recorded for carrots stored in lesser than 50 % O_2/30 % CO_2.

Passive MAP at both temperatures led to a severe decrease in pH of carrots. Inoculation with microorganisms such as *Listeria* or *Salmonella* affects the product's pH probably because of metabolism by-products. Treatment with chlorine dioxide induced minor pH changes versus storage (Fig. 1.1).

Ready-to-eat (RTE) carrots, after being peeled and sliced, were packed (4 °C) under passive (PP film used) and A MAP at low (5 % O_2, 10 % CO_2,) and high oxygen concentrations (80 % O_2, 10 % CO_2) after dipping into citric acid

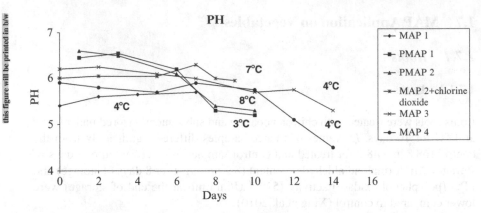

Fig. 1.1 Changes in pH of carrots under passive or active MAP vs. storage time [MAP 1 (2.1 % O_2/4.9 % $CO2$/93 % N_2), Tassou and Boziaris (2002), PMAP 1 (P-Plus with equilibrium atmospheric conditions of 10 % O_2/10 % CO_2), PMAP 2 (P-Plus with equilibrium atmospheric conditions of 6 % O_2/15 % CO_2), Barry-Ryan et al. (2000), MAP 2 (4.5 % O_2/8.9 % CO_2/86.6 % N_2) for grated carrots, Gomez-Lopez et al. (2007a), MAP 3 (2.1 % O_2/4.9 % CO_2/93 % N_2 and inoculation with *Salmonella* strains), MAP 4 (2.1 % O_2/4.9 % CO_2/93 % N_2 and inoculation with *L. monocytogenes* strains), Kakiomenou et al. (1998)]

(0.1 % w/v for 15 min). The deterioration of the specimens' texture at both passive and A MAP conditions especially after 14 days of storage suggested significant softening. However, there was neither yeast nor mould growth reported during the 21 days of storage in any of the applications (Ayhan et al. 2008).

Different packaging films (OPP, Pebax with hydrophilic coating, polyether block amide, P-plus 1, P-plus 2) and storage temperatures (3 and 8 °C) were applied to set up a range of EMA for storage of shredded carrots. The product packed in P-plus 1 at 3 °C and in OPP at 8 °C had the highest appearance scores, whereas the highest aroma scores were attributed to product packed in the P-plus films. A P-plus microporous film was found to be the most suitable for the storage of shredded carrots according to Barry-Ryan et al. (2000).

Shredded, dipped in a chlorine solution, inoculated with *L. monocytogenes* strains carrots, were stored (at 5 and 15 °C) under MAP (3 % O_2). Cutting method (whole or shredded carrots), chlorine additive, and MAP had no impact on the survival or growth of *L. monocytogenes* or naturally occurring microflora. Populations of mesophilic aerobes, psychrophiles, and yeasts and moulds enhanced throughout storage at 5 and 15 °C according to Beuchat and Brakett (1990).

Carrot discs at 4 and 8 °C with edible coating (Natureseal) prior to packaging (OPP or microperforated polypropylene PA_{60} film) were used by Cliffe-Byrnes and O'Beirne (2007) to allow for MA conditions for storage. The use of Natureseal had substantial beneficial effect on the appearance of carrot discs packaged with PA_{60} improving sensory scores, which appeared to be due to a reduction in surface whitening and moisture loss. No ethanol was detected in PA_{60} packages.

Gleeson and O'Beirne (2005) investigated the inoculation of sliced carrots (with blunt or sharp machine blade and a razor blade) with *E. coli* and *L. innocua* and storage under passive MAP (8 °C). They found that *E. coli* counts on razor sliced carrots were approximately 1 log cycle lower than counts on the blunt or sharp machine blade sliced carrots. Subsequently, *E. coli* survived better on carrots sliced with blades that caused the most damage to cut surfaces.

Gomez-Lopez et al. (2007a) monitored the effect of gaseous chlorine dioxide treatment in conjunction with MAP (4.5 % O_2 + 8.9 % CO_2) to reach shelf life extension of grated carrots stored at 7 °C. After ClO_2 treatment, the decontamination levels (log CFU/g) obtained were 1.88, 1.71, 2.60, and 0.66 for mesophilic aerobic bacteria, psychrotrophs, lactic acid bacteria (LAB) and yeasts respectively. The treated samples showed yeast counts that were initially only minimally reduced with respect to the untreated ones, and reached the same level on the fifth day (Table 1.1).

The survival of inoculated *S. enteritidis* and *L. monocytogenes* on shredded carrots under MAP (2.1 % O_2/4.9 % CO_2, 5.2 % O_2/5 % CO_2) at 4 °C was analyzed by Kakiomenou et al. (1998). Even though both pathogens survived they did not grow regardless of the packaging system used. It was detected that numbers of total viable counts (TVC) and LAB were always lower in samples stored under MA conditions than in those stored aerobically.

Grated carrots, inoculated with *E. coli* (6 log CFU/g) and stored at 4 ± 1 °C were used to assess the efficiency of gamma irradiation (doses from 0.15 to 0.9 kGy) in conjunction with MAP (60 % O_2 + 30 % CO_2) by Lacroix and Lafortune (2004). Inoculated *E. coli* was completely eliminated at doses ≥0.3 kGy in samples treated under MAP. A 4 and 3 log reduction in samples treated under MAP and under air was reported.

Carrots were washed in hydro-alcoholic solution (30 % v/v in ethanol), coated into a sodium alginic (4 % w/v) water solution and then washed again in the alcoholic solution and subsequently stored under both PMAP (PP film with OTR: 2,076.9 cm^3/m^2 day) and AMAP (10 % O_2/10 % CO_2) at 4 °C. Lag phase of microbial population on coated samples was extended with cell loads remaining stable at 3 log CFU/g for the initial 7 days. Combination of both PMAP and AMAP and coating of carrots offered a shelf life extension of 12 days compared to control (Mastromatteo et al. 2012).

Pilon et al. (2006) studied carrots stored under air, vacuum and MA conditions (2 % O_2 + 10 % CO_2) at 1 ± 1 °C for 21 days (LDPE-BOPP film). Average values recorded for pH ranged from 6.2 to 6.5 for minimally processed carrot, over different storage periods, whereas vitamin C remained intact up to 21 days of storage. For the minimally processed carrot, faecal and total coliforms, anaerobic mesophiles and *Salmonella* could not be detected in any of the treatments.

Grated carrots stored in air, MA (2.1 % O_2, 4.9 % CO_2) were used to monitor the fate of inoculated *Salmonella enteritidis* and vacuum at 4 °C to trace the presence of *Lactobacillus* irrespectively of the size of its inoculum. As Tassou and Boziaris (2002) pointed out, acetic acid was produced mostly under modified atmosphere in comparison with samples stored under vacuum or aerobic conditions.

Table 1.1 Brief summary of conditions used for carrot storage under MAP

Species and food type	Initial gas mix	Packaging material	Treatment before packaging	Storage temperature (°C) and storage period (days)	Color	Microflora	Texture-weight loss	Sensory analysis	Shelf life (days)-life extension	References
Carrot (*Daucus carota L.*)	PMAP	1. OPP with OTR: 3,500 cm^3 m^{-2} day^{-1} atm.$^{-1}$ 2. OPP with OTR: 1,100 cm^3 m^{-2} day^{-1} atm.$^{-1}$	Sterilization with 70 % ethanol prior to packaging. Inoculation with *E. coli* O157:H7 strains	5 and 25 °C —8 days		In all studied conditions *E. coli* strains managed to survive (3.2, 3.8 and 4 log CFU/g for samples stored in film 1, 2 and control samples, respectively at 5 °C) and in packages stored at 25 °C the population increased significantly (8, 7 and 8.4 log CFU/g for samples stored in film 1, 2 and control samples, respectively)			Survival of *E. coli* strains in all tested conditions stresses the need for effective disinfection measures of ready-to-eat vegetables	Abadias et al. (2012)

Sample	Atmosphere	Packaging	Disinfection/Treatment	Storage						Reference
Orange and purple carrots	1. 5% O$_2$/5% CO$_2$ 2. 95% O$_2$/5% CO$_2$	PE bags	Disinfection for 5 min (100 ppm free chlorine solution and subsequent shredding of the samples)	5 ± 2 °C—13 days	Browning was absent on orange carrots for the whole storage period while purple carrots under MAP 1 had better results compared to the other samples		Purple carrots stored under MAP 2 showed a sharp decrease of antioxidant activity	Sensory attributes of purple carrots were better preserved by MAP 1	Shelf life of purple carrots stored under MAP 1 was extended by 2 to 3 days compared to other samples	Alasalvar et al. (2005)
Sliced carrots cv. Amsterdamse bak	Control Atmosphere: 1. 90% O$_2$/10% CO$_2$ 2. 80% O$_2$/20% CO$_2$ 3. 50% O$_2$/30% CO$_2$ 4. 70% O$_2$/30% CO$_2$ 5. 1% O$_2$/10% CO$_2$	Airtight containers with continuous flush of combination of gases	1. Distilled water 2. NaOCl solution (200 mg/L), chlorine or 5% (v/v) H$_2$O$_2$, 3. 0.1 and 0.5% (w/v) citric acid 4. Alginate coating (S170 + 2% CaCl$_2$), combination of the last two treatments	8 °C—15 days		The combination of alginate coating and 0.5% citric acid treatment resulted in significant decrease of the microbial load (1.9, 2, 1.4 and 1.9 log CFU/g reductions on TVC, LAB, Pseudomonas, Entero-bacteriaceae populations, respectively compared to control samples)	Citric acid treatment and application of edible coating minimized texture changes for 8 days (855 and 828 N for samples treated with citric acid and the combination of both acid and coating, respectively)	Citric acid treatment combined with the use of edible coating resulting in preserving quality characteristics of carrots for 10 days. MAP 3 combined with citric acid pretreatment preserved sensory characteristics for 13 to 15 days of storage	MAP 3 resulted in prolonging shelf life of carrots by 2 to 3 days compared to control. The use of pretreatment resulted in prolonging shelf life by 5 to 7 days (13 to 15 days total)	Amanatidou et al. (2000)

(continued)

Table 1.1 (continued)

Species and food type	Initial gas mix	Packaging material	Treatment before packaging	Storage temperature (°C) and storage period (days)	Color	Microflora	Texture-weight loss	Sensory analysis	Shelf life (days)-life extension	References
Carrots (*Daucus carota*) cv. Nantes	1. 5 % O_2/10 % CO_2 2. 80 % O_2/10 % CO_2 3. PMAP	CPP-OPP with OTR and CDTR of 1,296 and 3,877 cm^3/ m^2 day, respectively	Dipping into citric acid solution (0.1 % w/v) for 15 min, and then sliced. The sliced carrots were dipped into citric acid solution (0.1 % w/v) for 10 min	4 °C—21 days	Color attributes were successfully retained on all conditions for the 21 days of storage	Significant differences were evident on mesophile growth with MAP 2-treated samples having the lowest microbial numbers (7.92, 8.25 and 7.48 log CFU/g for PMAP, MAP 1 and MAP 2, respectively on the 21st day)	Samples lost their initial texture properties on the 14th day of storage and onwards (73.61– 94.54 N on the 14th day for all storage conditions)	Samples stored under super-atmospheric conditions retained quality characteristics better compared to other storage conditions	PMAP and MAP 2-treated samples had a shelf life of 7 days, whereas MAP 1 treated samples had a limited shelf life of 2 days	Ayhan et al. (2008)

Product	Gas composition	Film	Treatment	Temperature	Microbial	Firmness	Appearance	Other	Reference
Carrots (*Daucus carota L.*) cv. Nantaise des Sables	Passive MAP: 1. >1 % O$_2$/30 % CO$_2$ 2. >1 % O$_2$/4 % CO$_2$ 3. >1 % O$_2$/16 % CO$_2$ 4. 10 % O$_2$/10 % CO$_2$ 5. 18 % O$_2$/3 % CO$_2$	1. OPP with OTR: 1,200 mL/ m^2 day atm. 2. Pebax film with OTR: 6,500 mL/ m^2 day atm. 3. Polyether block amide with OTR: 13,000 mL/ m^2 day atm. 4. P-plus 1 with OTR: 25,000 mL/ m^2 day atm. 5. P-plus 2 with OTR: 200,000 mL/ m^2 day atm.	Chlorine treatment (into water with 100 µg/mL free chlorine) for 5 min. Carrots were then shredded	3 and 8 °C—10 days	A 0.5 log CFU reduction on mesophile load was achieved with the use of P-plus 1 film from day 5 and onwards	Firmness values showed a gradual decrease from day 8 and onwards	Samples stored in P-Plus 1 bags had the highest appearance score followed by samples stored in OPP		Barry-Ryan et al. (2000)
Carrots (*Daucus carota*)	1. 3 % O$_2$/ 97 % CO$_2$	Film with OTR: 3,000 mL/ m^2 24 h atm.	Shredding, chlorine treatment (water with 200 µg/mL) and inoculation with 2 strains of *L. mono-cytogenes*	5 and 15 °C—24 and 7 days, respectively	Raw carrots posed an antimicrobial effect, reducing *L. monocytogenes* populations. Chlorine treatment effectively reduced microbial population of yeasts, moulds and mesophiles by 90 %			Cutting treatments, chlorine treatment and MAP did not show any significant effect in *L. monocytogenes* populations	Beuchat and Brakett (1990)

(continued)

Table 1.1 (continued)

Species and food type	Initial gas mix	Packaging material	Treatment before packaging	Storage temperature (°C) and storage period (days)	Color	Microflora	Texture-weight loss	Sensory analysis	Shelf life (days)-life extension	References
Carrots (*Daucus carota*) cv. Nairobi	Passive MAP: 1. 1 % O_2/12 % CO_2 at 4 °C 2. 1 % O_2/16 % CO_2 at 8 °C 3. 7 % O_2/12 % CO_2 at 4 °C 4. 4 % O_2/15 % CO_2 at 8 °C	1. Monoaxial OPP micro-perforated 35 µm 2. OPP film PA-60	Antimicrobial treatment with chlorine solution at 4 °C. Dipping in an edible coating, Natureseal solution for 30 s	4 °C or 8 °C—6 days	Carrots stored in PA-60 bags had higher $L*$ values		Texture character-istic were better retained with the use of PA-60	Sensory attributes were better maintained in PA-60 bags and with the use of Nature-seal edible coating. The almost anaerobic conditions created by OPP film led to the incidence of off-odors and tissue softening		Cliffe-Byrnes and O'Beirne (2007)

Product	EMA	Packaging	Treatment	Storage	Results	Sensory	Conclusion	Reference
Irish carrots (cultivar Nairobi)		OPP 35 μm	Carrots were cut with: 1. Blunt machine blade 2. Sharp machine blade 3. Razor blade Inoculation with *E. coli* and *L. innocua* strains	8 °C for 9 days	The slicing methods that caused increased damage on the cut surfaces induced microbial growth. Samples cut with a razor blade had lower *E. coli* numbers (1 log CFU difference) from both mechanical cutting methods. Mesophile numbers showed no significant differences		Slicing using a mild technique that does not cause increased tissue damage lowers the possibility of pathogens' survival and growth	Gleeson and O'Beirne (2005)
Grated carrots (*Daucus carota L.*)	EMA: 4.5 % O_2/8.9 % CO_2	Film with OTR: 3,529 mL O_2/kg h	ClO_2 gas treatment	7 °C—9 days	ClO_2 pretreatment significantly reduced microbial load (1.88, 1.71, 2.6 and 0.66 log CFU/g reduction of APC, psychrotrophs, LAB and yeasts, respectively). Growth of APC, psychrotrophs and LAB was delayed by 2 days	Sensory attributes were not significantly affected by ClO_2 pretreatment. The factor that led to sensory rejection of treated samples was off-odors formation	Shelf life extension achieved by gaseous chlorine dioxide pretreatment was one day due to increased yeast contamination	Gomez-Lopez et al. (2007a, b)

(continued)

Table 1.1 (continued)

Species and food type	Initial gas mix	Packaging material	Treatment before packaging	Storage temperature (°C) and storage period (days)	Color	Microflora	Texture-weight loss	Sensory analysis	Shelf life (days)-life extension	References
Shredded carrots (*Daucus carota L.*)	AMAP: 1. 2.1 % O$_2$/4.9 % CO$_2$ 2. 5.2 % O$_2$/5 % CO$_2$ 3. PMAP	PE bags 60 µm thick with OTR: 3,841 mL/ m^2 24 h atm. and CDTR: 33,875 mL/ m^2 24 h atm. at 20 °C	Inoculation with *S. enteritidis* and *L. mono-cytogenes* strains	4 °C ± 0.2 °C —14 days		*S. enteritidis* and *L. monocytogenes* population survived but significantly reduced during storage under all storage conditions (1.3 and 1.7 log CFU/g reductions of *Salmonella* population for control and MAP 1 samples, respectively). TVC and LAB counts were always lower on samples stored under MAP A compared to air samples (PMAP)			Growth of *Listeria* was significantly affected by both atmosphere composition, pH of the in package environment, vegetable type and competition with the existing flora	Kakiomenou et al. (1998)

Grated Carrots (*Daucus carota*)	1. PMAP 2. 60 % O_2/30 % CO_2	0.5 mm metallized polyester/2 mm EVA copolymer sterile bag with OTR: 0.7 mL/ $m^2 \cdot$ day	Gamma irradiation at a dose of 0.15, 0.3, 0.6, 0.9 kGy. Inoculation with *Escherichia coli* (10^6 CFU/g)	4 ± 1 °C—50 days	Irradiation (0.15 kGy dose) and subsequent AMAP or PMAP storage led to a 3 and 4 log reduction of *E. coli* population on day 1 for PMAP and AMAP samples, respectively, but a 3 log CFU/g population was detected on day 7. Doses ≥ 3 kGy and AMAP storage led to complete eradication of *E. coli* for the whole storage period of 50 days		Disinfection with irradiation and storage under AMAP are a useful combination for preserving quality of carrots	Lacroix and Lafortune (2004)

(continued)

Table 1.1 (continued)

Species and food type	Initial gas mix	Packaging material	Treatment before packaging	Storage temperature (°C) and storage period (days)	Color	Microflora	Texture-weight loss	Sensory analysis	Shelf life (days)-life extension	References
Carrots (*Daucus carota L.*)	1. PMAP 2. AMAP: 10 % O_2/10 % CO_2	Microperforated PP film (30 μm thick) with OTR: 2,076.9 cm^3/m^2 day and CDTR: 1,924.6 cm^3/m^2 day	Dipping into a hydro-alcoholic solution (30 % v/v in ethanol). Coating into a sodium alginic (4 % w/v) water solution and subsequent washing into alcoholic solution	4 °C—19 days		Mesophile load of coated samples in PMAP and AMAP remained stable (3 log CFU/g) until the seventh day and then increased up to 5.6 log CFU/g. Psychrophile load of coated samples was lower under AMAP compared to PMAP (4 and 5 log CFU/g for AMAP and PMAP samples, respectively, at the end of storage period). No Enterobacteriaceae load was detected on coated samples under AMAP			Shelf life of uncoated samples was limited to 2 days due to poor sensory attributes. Shelf life of coated samples stored under AMAP and PMAP was 14.99 and 13.87 days, respectively	Mastromatteo et al. (2012)

Product	Treatment	Package	Process	Storage	β-carotene	Microbiology	Weight loss	Vitamin C / pH	Reference
Carrots (*Daucus carota L.*) cv Nantes	1. Vacuum 2. Active MAP: 2 % O_2/10 % CO_2 3. PMAP	BOPP/LDPE bags	Immersion in cold water (7 °C) with 100 mg/L of free chlorine at pH 7.0 for 15 min. Carrots were cut into cubes	1 °C ± 1 °C —21 days	β-carotene decreased slightly during storage	Total and fecal coliforms, anaerobic mesophiles and *Salmonella* were absent in carrots stored under all studied conditions. Psychrotroph counts remained ≤4 log CFU/g for the whole storage period under all storage conditions		Vitamin C was retained in all studied storage conditions for the whole experiment period (130.3–132.8 mg/100 g)	Pilon et al. (2006)
Grated carrots	1. PMAP 2. AMAP: 2.1 % O_2/4.9 % CO_2	PE bags OTR: 1,000 and CDTR: 5,450 mL/ m² day bar	Immersion in alcohol, flamed, peeled, again immersed in alcohol and flamed and grated with a sterile grater. The carrots were inoculated with *S. enteritidis* and *Lactobacillus* sp.	4 °C—12 days		*S. enteritidis* survived under all treatments being unaffected by the presence of *Lactobacillus*. LAB also grew in all studied conditions. The acidic environment created by LAB population played an inhibitory role in *Salmonella* growth	Weight loss of coated samples did not exceed 3 % until the end of the experiment	pH significantly dropped in all samples inoculated with *Lactobacillus*	Tassou and Boziaris (2002)

1.7.1.3 Ginseng

Passive MAP (PVC and LDPE films) at 0, 10 and 20 °C was applied to store fresh ginseng. The lowest weight loss was recorded at 0 °C (1.9 and 4.9 % for PVC and LDPE, respectively) and the highest weight loss at 20 °C (5.6 and 6.6 % for PVC and LDPE, respectively). Storage at 0 °C, especially in PVC packages helped to obtain lower decay rate (1 %) while at the same time its reducing sugar content increased from initial 8.2 to 22, 23.2 and 39.9 mg/g at 0, 10 and 20 °C respectively. The shelf life of ginseng was prolonged up to 210 days (Hu et al. 2005).

American ginseng roots with an antimicrobial agent (dipping in 0.5 % DF-100 solution for 5 min), were stored at 2 °C under CA (2 % O_2 and 2, 5, and 8 % CO_2) or passive MA [PD-941 (high), PD-961 (medium) and PD-900 (low permeability) polyolefin films used] conditions and studied by Jeon and Lee (1999). Ginseng samples stored under CA at 5 % CO_2 showed the least change in individual and total ginsenosides. No changes in free sugars (17.74 % total sugars at harvest and 16.48 and 19.34 for CA at 5 % CO_2 and PD-961, respectively) were recorded after 3 months.

Macura et al. (2001) studied the survival ratio of inoculated C. botulinum in MA (Winpac medium transmission laminated film used) ginseng roots stored at 2, 10, and 21 °C. The development of anaerobic conditions was more rapid at higher temperatures. The detection of C. botulinum toxin production at 10 °C was detected after 14 weeks of storage, before all products were spoiled and rendered unfit for human consumption. At 21 °C, the product spoiled first and then it became toxic.

1.7.2   Tubers

1.7.2.1   Sweet Potato

Sweet potatoes were sliced and stored under MA conditions using a low (PD-900), medium (PD-961) and high (PD-941 polyolefin) permeability film at 2 °C or 8 °C for 14 days. Samples stored in PD-900 and PD-961 films had significantly lower weight losses compared to those in PD-941 film bags. Higher storage temperature doubled the weight loss compared to samples stored at 2 °C. B-carotene content of fresh-cut sweet potatoes at the end of storage had an average of 6.01 mg per 100 g of FW (Erturk and Picha 2007).

McConnell et al. (2005) packed shredded sweet potatoes from two major commercial cultivars ('Beauregard' and 'Hernandez') in low (PD-900) and medium (PD-961) O_2 permeability bags at 4 °C and flushed them with gas composed of 5 % O_2 and 4 % CO_2. Samples from 'Beauregard' and 'Hernandez' cultivars stored under MAP did not undergo any change in ascorbic acid contents (15.4 and 14.7 mg/100 g, respectively). Sweet potatoes in PD-900 and PD-961 exhibited more than one log lower total aerobic counts compared to control and PD-941 samples (7 and 8.2 log CFU/g, respectively) for 'Beauregard' cultivar.

1.7.2.2 Potato

In another research work, the effect of γ-irradiation dose (0–1.5 kGy), citric acid (0–1.0 %) and potassium metasulphite (KMS) dipping solutions concentration (0–1.0 %) on potato cubes stored under MAP (PP film) at 4 ± 1 °C for 4 weeks was studied. At the end of the experiment the best quality parameters (L^* value was ≥ 48.50 %, a^* value ≤ 0.95, b^* value ≤ 7.5, hardness ≥ 100 N, sucrose concentration ≤ 0.19 % and sensory score ≥ 6.0) obtained for samples irradiated (dose 1.0 kGy) and dipped (citric acid concentration 0.33 % and KMS concentration 0.55 %) prior to MAP (Baskaran et al. 2007).

The effectiveness of sanitizers [sodium sulphite (SS), sodium hypochlorite (SH), Tsunami (T), ozone (O_3) and the combination of ozone–Tsunami (O_3T)] on the sensory and microbial quality of fresh-cut potatoes stored under PMAP (LDPE film) and vacuum packaging (VP) at 4 °C was investigated by Beltran et al. (2005b). Browning was effectively controlled with maintenance of initial appearance only for samples dipped in SS. The combination of O_3 and T was found to be the most effective in reducing LAB, coliforms and anaerobic bacteria numbers by 3.3, 3.0 and 1.2 logs, respectively.

Gunes and Lee (1997) carried out an experiment within the frame of which potatoes were peeled with the abrasive peeler, a hand-peeler and a lye solution, treated with antibrowning agents (dipping in 0.5 % L-cysteine, 0.5 % L-cysteine plus 2 % citric acid mixture, 5 % ascorbic acid, 0.1 % potassium metabisulfite and stored under AMAP with 9 % CO_2/3 % O_2) and stored under active (9 % CO_2/3 % O_2, 9 % CO_2 and 100 % N_2 compositions) MAP at 2 °C. The incidence of browning was reduced with the use of cysteine-citric acid mixture (6 % reduction of L-value). Manual peeling proved to be the best method with the lowest L^*-value reduction (3 %).

1.7.3 Leaves

1.7.3.1 Lettuce

The microbial quality of '*Lollo Rosso*' lettuce under MAP at 5 °C with initial gas atmosphere of 3 kPa O_2 and 5 kPa CO_2 was examined by Allende et al (2004b) and showed that sensory attributes limited shelf life to 7 days due to severe off-odors. LAB enhanced from 3 to 6 log CFU/g after 7 days of storage and for mesophiles and psychrophiles there was a 3 log increase (from 5 to 8 log CFU/g) for the same period. Subsequently, the sensory evaluation improved and tissue browning was reduced due to the presence of low oxygen atmosphere.

Ares et al (2008) investigated the effect of both passive (PP film) and active (initial gas mixture of 5 % O_2 and 2.5 % CO_2) MAP on sensory shelf life of butterhead lettuce leaves, stored at 5 and 10 °C. Lettuce leaves in PMAP, packaged and stored at 5 °C, resulted in considerably higher off-odor, wilting appearance, and presence of stains on the leaf surface or browning on the midribs than those in AMAP, thus indicating a higher rate of deterioration.

Combined ultraviolet-C radiation (0.4, 0.81, 2.44, 4.07, and 8.14 kJ/m^2) and passive MAP (BOPP film used) were applied by Allende and Artes (2003a) to diminish the growth of psychrotrophic (0.5–2 log CFU/g reduction) and coliform bacteria as well as yeast growth. The sensory quality of lettuce pieces was not adversely affected by this combined treatment when compared to control samples.

Aerobic counts with microbial limit of 7 log CFU/g were established by the Spanish legislation exceeded on day 5 for control, and by day 8 for product treated with 4.06 or 8.14 kJ/m^2 thereby showing that UV-C treatment and storage at 5 °C under MAP (bioriented polypropylene film (BOPP) used) conditions prolonged shelf life of 'Lollo Rosso' lettuce. Enhanced tissue brightness was due to the highest dose whereas browning was reduced when 2.44, 4.07, and 8.14 kJ/m^2 were applied (Allende and Artes 2003b).

Iceberg lettuce was examined in regard with the impact of packaging atmosphere (PMAP with OPP film used with or without N_2 flushing) and storage temperature (3 and 8 °C) on total ascorbic acid (TAA) by Barry-Ryan and O'Beirne (1999). TAA retention (14 and 12 mg/100 g for nitrogen flushed and OPP bags, respectively) over that in unsealed bags (11 mg/100 g) increased by sealing the OPP bags (passive MAP or nitrogen flush) and allowing a MA to develop.

AMAP (4 kPa O_2 and 12 kPa CO_2), as described by Beltran et al. (2005a) was effective in controlling microbial growth in iceberg lettuce, thereby reaching 2.0 log reduction in relation to samples stored in air. AMAP initiated a 2.0–3.5 log reduction of coliforms on sanitized samples [washing at 4 °C using three ozonated water dips (10, 20, and 10 mg L^{-1} activated with ultraviolet-C light minimum total ozone dose)], compared with water-washed samples. Shredded lettuce had an excellent visual quality after washing for all treatments, and no browning was observed.

Two types of atmosphere modification (PMAP with PE bags used and A MAP (AMAP) with the same bags and initial atmosphere 10 % O_2–10 % CO_2) and initial wash with chlorinated water was tested on romaine lettuce stored at 8 and 20 °C. The treatment with chlorinated water reduced initial mesophile counts by 1.5 log CFU/g.

Bidawid et al. (2001) investigated the effect of various modified atmospheres (CO_2, N_2 at 30:70, 50:50, 70:30 and 100 % CO_2) on the survival rate of hepatitis A virus (HAV) on romaine lettuce stored at 4 °C and at room temperature. Lettuce stored under a MA of 70 % CO_2 was found to have the highest HAV survival rates (83.6 %). The virus survival rates diminished significantly with an increase at storage temperature. The higher CO_2 level caused injury and resulted in enhanced browning, particularly under 100 % CO_2.

Chua et al (2008) undertook the effort to evaluate the importance of temperature abuse over the storage period. Several enterohemorrhagic E. coli strains, in spite of their dysfunctional rpoS locus, were able to profit from MAP conditions and metabolites presence from lettuce surfaces to induce acid resistance at growth permissive temperatures (≥ 15 °C). However, for MAP-stored lettuce at temperatures ≤ 10 °C or for lettuce under aerobic conditions, no acid resistance was induced. Low storage temperature and co-existence with L. innocua are hurdles for the growth of

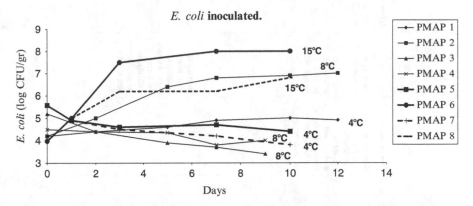

Fig. 1.2 *E. coli* colonies inoculated on lettuce and stored under modified atmospheres. [PMAP 1 (with steady-state atmosphere of 3–4 % O_2/10–12 % CO_2), for iceberg lettuce, PMAP 2 (with steady-state atmosphere of 3–4 % O_2/10–12 % CO_2) for iceberg lettuce, Francis and O'Beirne (2001), PMAP 3 (with oriented polypropylene as a packaging film) and inoculation with *E. Coli* and *L. innocua* for razor blade cut iceberg lettuce, PMAP 4 (with oriented polypropylene as a packaging film) and inoculation with *E. coli* and *L. innocua* for razor blade cut butterhead lettuce, Gleeson and O'Beirne (2005), PMAP 5 and PMAP 6 (2 % O_2–98 % N_2 and a film with OTR: 110 mL/m^2 cm^3 day) and inoculation with *E. coli O157:H7* strains and PMAP 7 and PMAP 8 (film with OTR: 110 mL/m^2 cm^3 day and with 40 microperforations) and inoculation with *E. coli O157:H7* strains (Sharma et al. 2011)]

inoculated *E. coli*, limiting its population down to acceptable levels, as can be seen in Fig. 1.2.

An investigation was also carried out on the effect of different O_2 levels from 0 to 100 kPa in conjunction with 0, 10 and 20 kPa CO_2 on the respiration metabolism of greenhouse grown fresh-cut butter lettuce. The concluding remark is that 80 kPa O_2 (superatmospheric conditions) must be used in MAP to prevent fermentation of lettuce when combined with 10–20 kPa CO_2 for the reduction of their respiration rate (Escalona et al. 2006).

Fan and Sokorai (2011) investigated the effect of different irradiation doses (0.5 and 1 kGy) on quality of iceberg lettuce stored under MAP (nitrogen gas flush and storage in bags with 6,000–8,000 mL/m^2/day OTR) at 4 °C for 14 days. The irradiated samples stored under MAP maintained quality over the limit of sales appeal (the limit was set to 5), while those stored at air showed extended tissue browning and quality degradation (Table 1.2).

Treatment with warm water (dipping in water for 2 min in either 5 or 47 °C) before irradiation (0, 0.5, 1 or 2 kGy) of iceberg lettuce and packaging under passive MAP (films with OTR: 4,000 mL/h m^2) at 3 °C was studied by Fan et al. (2003). Less tissue browning was reported for lettuce dipped at 47 °C and irradiated at 0.5 and 1 kGy in comparison to corresponding irradiated samples dipped at 5 °C. Irradiated lettuce had better overall physical and sensory properties than non-irradiated lettuce dipped at either 47 or 5 °C. Irradiation induced higher antioxidant content which was nulled by the 47 °C water dip effect.

Table 1.2 Collected data from researches on lettuce stored under MAP

Species and food type	Initial gas mix	Packaging material	Treatment before packaging	Storage temperature (°C) and storage period (days)	Color	Microflora	Texture-weight loss	Sensory analysis	Shelf life(days)-Life extension	References
'Red Oak Leaf' lettuce (*Lactuca sativa L.*)	(1) 2–10 % O_2/5–12 % CO_2	BOPP with OTR: 1,800 mL/m^2 day atm.	UV-C treatment with (1) 0.41, (2) 0.81, (3) 2.44, (4) 4.07, and (5) 8.14 corresponding doses (kJ/m^2)	5 °C—9 days	Browning was not intense on UV treated samples compared to control	Microbial population was severely inhibited by MAP and UV-C treatment combination (0.5–2 and 0.7 log CFU/g reduction of psychrophiles and yeasts respectively, compared to control) while the reduction was linearly related to UV-C dosage. LAB growth was not affected		Sensorial attributes were not significantly affected by treatment and atmosphere change due to respiration rate increase	Shelf life was estimated to be 7 days with no significant differences between treatments	Allende and Artes (2003a)
'Lollo Rosso' lettuce (*Lactuca sativa.*)	(1) 2–10 % O_2/5–12 % CO_2	BOPP with OTR: 1,800 mL/m^2 day atm.	UV-C treatment with (1) 0.41, (2) 0.81, (3) 2.44, (4) 4.07, and (5) 8.14 corresponding doses (kJ/m^2)	5 °C—10 days	High UV doses significantly decreased tissue browning (2.44, 4.07, and 8.14 kJ/m^2)	Only the highest UV-C dose significantly decreased psychrotrophic bacteria, coliform, and yeast growth (>1 log CFU difference from control for psychrotrophs, 1 log CFU/g difference for coliforms). LAB growth was intense maybe due to inhibition of competitive flora			Shelf life of UV-C treated samples (4.07, and 8.14 kJ/m^2) was 8 days while control samples had a shelf life of 5 days	Allende and Artes (2003b)

Product	MAP conditions	Packaging	Pre-treatment	Storage	Appearance	Microbial	Weight loss	Sensory	Shelf life	Reference
Red pigmented lettuce (*Lactuca sativa*, "Lollo Rosso")	3 % O_2/5 % CO_2 active MAP	Bags of 35-μm PP film	Lettuce was washed with chlorinated water (160–180 ppm) and shredded prior to packaging	5 °C—7 days		Washing decreased microbial numbers (3 log reduction for psychrotrophs and LAB and 2 log reduction for coliforms). Microbial shelf life of MAP treated samples was 6 days (8 log CFU/g for mesophiles and psychrophiles and 6 log CFU/g for LAB)	Weight loss was minimal (0.28 % of total weight) after 7 days	Samples under MAP were sensorially unacceptable after 7 days of storage	Shelf life of MAP treated samples stored at 5 °C was limited to 6 days by both microbial proliferation and sensory rejection	Allende et al. (2004b)
Butterhead lettuce (*Lactuca sativa L.*, cv Wang)	1. Passive MAP 2. Active MAP with 5 % O_2 and 2.5 % CO_2	PP films (40 μm thickness)	Chlorinated water (200 ppm total chlorine) for 10 min	5 and 10 °C for 49 and 21 days respectively	Wilting was more intense on samples stored at 10 °C compared to those stored at 5 °C		Weight loss on samples from all storage conditions was over 20 % at the end of the storage period	Sensory attributes were preserved better on samples stored under AMAP at 5 °C. Dark stains appeared on the surface of the samples after 37, 46, 17 and 15 days for samples under AMAP and PMAP stored at 5 and 10 °C, respectively	Shelf life was 43, 36, 14 and 14 days for samples under AMAP and PMAP stored at 5 and 10 °C, respectively due to unacceptable browning of the surface	Ares et al. (2008)
Spanish Iceberg lettuce (cultivar Salodin)	1. Passive MAP 2. Nitrogen flush (100 % N_2)	35 μm thick OPP with OTR: 1,200 mL/ m² day atm. and CO_2TR: 4,000 mL/m² day atm.	Samples were shredded into 6 mm wide pieces, either manually or mechanically. Samples were subsequently dipped for 5 min in a 100 ppm chlorine solution	3 and 8 °C—10 days	PMAP samples had increased browning from day 6 and onwards			Sensory characteristics were better on samples manually processed and stored under MAP with N_2 flushing	MAP, Nitrogen flushing, manual tearing and storage at 3 °C was the best combination for ascorbic acid retention	Barry-Ryan and O'Beirne (1999)

(continued)

Table 1.2 (continued)

Species and food type	Initial gas mix	Packaging material	Treatment before packaging	Storage temperature (°C) and storage period (days)	Color	Microflora	Texture-weight loss	Sensory analysis	Shelf life(days)-Life extension	References
Iceberg lettuce (*Lactuca Sativa L.*)	1. 4 % O_2/12 % CO_2 AMAP	PET—PP multilayer film with OTR: 4.2×10^{-13} mol s^{-1} m^{-2} Pa^{-1}	Three ozonated water dips were used [10, 20, and 10 activated by ultraviolet C (UV-C) light mg/L min total ozone dose], and were compared with chlorine rinses (80 mg/L).	4 °C—13 days	Browning was delayed on samples washed with chlorine	Mesophile load was reduced by 1.6 and 2.1 log CFU/g on ozone and chlorine-treated samples compared to control. A 2 log CFU/g reduction was achieved by AMAP. Coliform growth was inhibited on sanitized samples stored under AMAP by 2–3.5 log CFU/g compared to control		Sensory attributes of treated samples under MAP were not severely affected, being acceptable after 13 storage days	Ozonated water combined with MAP are treatments that preserved lettuce sensory quality and antioxidant constituents	Beltran et al. (2005a)
Romaine lettuce	1. Passive MAP 2. 30 % CO_2 3. 50 % CO_2 4. 70 % CO_2 5. 100 % CO_2	Barrier plastic bags of OTR (0.46– 0.93 mL/100 mL day atm.)	Inoculation of lettuce with *HAV* was performed by spreading 10 μL of virus-containing solution (1.7×10^5 plaque forming unit)	4 °C—12 days	Browning was more evident on samples under 100 % CO_2	*HAV* survival rate was higher on high CO_2 MAP atmospheres (>80 %) while the lowest rates were on samples stored on Petri dishes at air (47.5 %)				Bidawid et al. (2001)
Romaine lettuce (*Lactuca Sativa L.*)	Active MAP with evacuation until O_2 is 1 % and 4 % CO_2	Barrier film with OTR: 16.6 pmol/s/m²/Pa	The leaves were washed in 100 mg/mL NaOCl solution and inoculated with 5 strains of hemorrhagic *E. Coli* (O157:H7, O26:H11, O55:H7, O91:H21, O111:H12)	5, 10, 15, 20 °C—7 days		All *E. coli* strains developed gastric acid resistance on samples stored at temperatures >15 °C. On the other hand no resistance was detected on strains stored at temperatures <10 °C				Chua et al. (2008)

Product	Packaging/MAP	Film	Temp—days	Results	Results	Results	Results	Reference
Butterhead lettuces (*Lactuca sativa L.*) cv. Zendria	Active CA with 0–100 % O_2 and 0, 10, 20 % CO_2		1, 5 and 9 °C—10 days	Respiration rate was reduced in moderate CO_2 storage conditions with increased O_2 levels		Fermentation of lettuce was avoided in packages with superatmospheric conditions (80 kPa O_2 and 10–20 CO_2)		Escalona et al. (2006)
Iceberg lettuce cv. Sharpshooter	PMAP	Film bags (E-300, Cryovac) with an OTR of 4,000 $cm^3/h/m^2$	3 °C—21 days	Irradiated and heat treated samples had better visual quality compared to control	Texture was better preserved on samples irradiated with low doses whereas high doses (2 KGy) increased cellular leakage and sogginess	Irradiation with low doses (0.5 and 1 KGy) and warm water treatment preserved sensorial characteristics better than all the other treatments without any losses in vitamin C or total antioxidants	Samples irradiated (0.5 and 1 KGy) heat treated and stored under PMAP remained sensorially acceptable for 21 days	Fan et al. (2003)
Iceberg lettuce (*Lactuca Sativa L.*) var Alladin	1. Passive MAP 2. Active MAP after gas flush with N_2. The resulting atmosphere was 4 % O_2–96 % N_2	35 µm OPP	3 and 8 °C—14 days	Storage at 8 °C of samples that underwent the antimicrobial dip resulted in significant increase of both *L. innocua* and *L. monocytogenes* populations by 2 log CFU/g. Mesophile bacteria reduction by the antimicrobial dip was 0.5–1 log CFU/g initially and reached 1–2 log CFU/g during the storage period			Both N_2 flushing and the use of antimicrobial dips favored *Listeria* survival at both storage temperatures while on samples stored at 8 °C the strains demonstrated and a significant growth	Francis and O'Beirne (1997)

Lettuce was dipped in either 5 or 47 °C water for 2 min, was packaged and then exposed to 0, 0.5, 1, or 2 kGy γ-radiation

Shredded lettuce was dipped in chlorine solution (100 ppm for 5 min) or citric acid (1 %, 5 min dip) solution. Then, the leaves were inoculated with *L. Innocua* and *L. monocytogenes* strains

(continued)

Table 1.2 (continued)

Species and food type	Initial gas mix	Packaging material	Treatment before packaging	Storage temperature (°C) and storage period (days)	Color	Microflora	Texture-weight loss	Sensory analysis	Shelf life(days)-Life extension	References
Irish iceberg lettuce	Passive MAP: 3–4 % O₂ and 10–12 % CO₂	35 μm OPP with OTR: 1,200 mL/m²/day/atm. and CO₂TR: 4,000 mL/m²/day/ atm.	Inoculation with L. monocytogenes and two nontoxigenic E. coli O157:H7 strains	4 and 8 °C—12 days		Pathogens populations increased (1.5 to 2.5 log CFU/g) on samples stored at 8 °C while during storage at 4 °C microbial load remained close to their initial numbers			L. monocytogenes and E. coli O157:H7 populations were reduced by reducing storage temperature from 8 to 4 °C	Francis and O'Beirne (2001)
Irish iceberg lettuce			Inoculation with 4 strains of L. monocytogenes with or without genes for glutamase decarboxylase resistance mechanism	4, 8 and 15 °C		The wild type L. monocytogenes strain showed higher survival rates compared to double mutant ΔgadAB				Francis et al. (2007)
Irish butterhead and iceberg lettuce	AMAP was created by N₂ flush prior to packaging	OPP 35 μm	Lettuce was: 1. Hand torn 2. Cut with blunt knife 3. Cut with razor blade And inoculated with E. coli and L. innocua strains	8 °C for 9 days		For butterhead lettuce razor sliced samples had significantly lower L. innocua numbers (by 1.2 log CFU/g) compared to hand torn ones or the samples cut with a blunt knife. On iceberg lettuce populations on razor sliced samples were higher compared to the other two processing methods (0.5 log CFU/g difference)			Mild processing techniques such as hand tearing or the use of a razor blade can result in reduced survival and growth of E. coli and L. innocua	Gleeson and O'Beirne (2005)

Product	MAP type	Film	Treatment	Storage	Browning	Microbial	Texture	Visual/sensory	Shelf life	Reference
Iceberg lettuce (*Lactuca sativa* L. capitata L.)	Passive MAP	Film with OTR: 3,529 mL O_2/m² 24 h atm.	Intense light pulse (ILP) treatment prior to packaging	7 °C for 9 days	Browning was intense on day 5 and onwards for treated lettuce samples	A 0.46 log CFU/g reduction on psychrotrophic counts was achieved by ILP. Yeasts counts remained at low levels with no significant differences from the control		Treated samples were rejected on day 3 due to poor visual quality	Both samples, treated and control had a shelf life of 3 days	Gomez-Lopez et al. (2005)
Iceberg lettuce (*Lactuca sativa* L. capitata L.)	2–4 % O_2 and 9 % CO_2	Film with OTR: 1,810 mL O_2/m² day atm.	Immersion for 1 min in a solution of 0.5 % HCl-L-cysteine monohydrate and treatment with ClO_2 (1.74 mg/L concentration)	7 °C for 9 days	Tissue browning was prevented by immersion in the L-cysteine solution	The reduction achieved by ClO_2 treatment was 0.84, 0.88 and 0.64 log CFU/g for APC, psychrotrophs and yeasts compared to control		Even though browning was significantly reduced by cysteine treatment, off odors and bad flavor were the limiting factors for treated samples	Shelf life of all samples, treated and untreated was limited to 4 days due to poor overall sensory quality	Gomez-Lopez et al. (2008)
Iceberg lettuce (*Lactuca Sativa L.*) var. Raleigh–Patriot	PMAP	PO laminate 44 µm thick with O_2 and CO_2 permeabilities of 3,800 and 13,000 mL m⁻² day⁻¹ atm.⁻¹	Lettuce was washed with chlorinated water (0.8–2.0 ppm of free chlorine) and γ-irradiated with 0.1–0.5 KGy 2 days after packaging	2±2 °C—10 days		Combination of both chlorine and irradiation treatment reduced both APC and yeast populations (290 CFU/g and 60 CFU/g, compared 220 000 and 1 400 CFU/g for treated and control samples, respectively)	Shear force of irradiated samples was significantly reduced compared to control		Shelf life of irradiated and control samples did not differ significantly	Hagenmaier and Baker (1997)

(continued)

Table 1.2 (continued)

Species and food type	Initial gas mix	Packaging material	Treatment before packaging	Storage temperature (°C) and storage period (days)	Color	Microflora	Texture-weight loss	Sensory analysis	Shelf life(days)-Life extension	References
Mixed lettuce (20 % endive, 20 % curled endive, 20 % radicchio lettuce, 20 % lollo rosso and 20 % lollo bionta lettuces—red and green variety)	AMAP with 3 % O_2/5 % CO_2	Bags with OTR: 1.04×10^{-11} mol O_2/m^2 s Pa	Inoculation with *L. monocytogenes* strains and *Aer. caviae*	2, 4, 7 and 10 °C for 11 days		Populations of both strains significantly increased on samples stored at elevated temperatures (0.58 and 1.29 log CFU/g outgrowth for *L. monocytogenes* and 0.65 and 0.73 log CFU/g for *Aer. caviae* at 7 and 10 °C, respectively)		Color and texture were the organoleptic properties that determined shelf life of samples	Shelf life of samples stored at 2, 4, 7 and 10 °C was 9, 7, 5 and 3 days, respectively	Jacxsens et al. (2002a)
Mixed lettuce (as mentioned above)	AMAP: 3 % O_2/5 % CO_2	40 µm bags with OTR: 2,026 mL O_2/m^2 24 h atm.	Inoculation with *L. monocytogenes* strains and *Aer. caviae*	T < 12 °C, t = 4 °C for 24 h, t = 5 °C for 2 h, t = 10 °C for 24 h, t = 5 °C for 2 h, t = 10 °C for 8 h, t = 7 °C for 48 h, t = 20 °C for 2 h and t = 7 °C for the rest of the experiment (7 days)	Browning on endive and lollo bionta leaves was severe and led to the final rejection	A sharp increase of spoilage microorganisms was detected after purchase and transportation to the domestic refrigeration (7.5, 5.8 and 5.2 log CFU/g for APC, LAB and yeasts, respectively). *L. monocytogenes* survived on samples while *Aer. caviae* managed to grow (0.58 log CFU/g day growth rate)		Off odor and poor color were the main reasons for sensorial rejection of the samples on the fifth day	Shelf life of samples on behalf of sensory attributes was 5 days, whereas on behalf of microbiological characteristics 4 days	Jacxsens et al. (2002b)

Product	MAP conditions	Packaging	Treatment	Storage	Microbial results	Sensory results	Shelf life	Reference
Mixed lettuce (as mentioned above)	AMAP: 3 % O_2 and 2–5 % CO_2	1. BOPP film (30 μm), PVC coated with an OTR of 15 mL O_2/m² 24 h atm. 2. High permeable packaging film with OTR: 2,270 mL O_2/m² 24 h atm. (EMA)		7 °C for 13 days	Psychrophile load reached rejection limit (8 log CFU/g) on the sixth day of storage for samples stored in bags from the second film while samples on BOPP bags had a psychrophile load of 7.2 log CFU/g on the same day. Yeast rejection limit was exceeded on the sixth day for samples stored under EMA (5.2 log CFU/g)	Musty taste, poor odor and color attributes were the limiting sensory factor that led to samples stored under EMA rejection on the sixth day	Shelf life of samples stored under EMA based on microbial count was 6–10 days for TPC and 6 days for yeasts and based on sensory quality was limited to 6 days. Samples' stored in BOPP bags shelf life was 4 days	Jacxsens et al. (2003)
Lettuce	AMAP 1.MAP A:21 % O_2/4.9 % CO_2 2. MAP B: 5.2 % O_2/5 % CO_2	PE bags with OTR 3,841 mL O_2/m² 24 h bar at 20 °C	Inoculation with S. enteritidis and L. monocytogenes strains	4 °C ± 0.2 °C for 14 days	Both pathogens survived but did not grow on both studied temperatures. Microbial populations varied on both studied samples (8, 7.6 and 5.2 log CFU/g for TVC, LAB and S. enteritidis, respectively on the 14th day on samples inoculated with Salmonella strains and 7.9, 6.8 and 4.5 log CFU/g for TVC, LAB and L. monocytogenes for samples inoculated with Listeria strains)			Kakiomenou et al. (1998)

(continued)

Table 1.2 (continued)

Species and food type	Initial gas mix	Treatment before packaging	Packaging material	Storage temperature (°C) and storage period (days)	Color	Microflora	Texture-weight loss	Sensory analysis	Shelf life(days)-Life extension	References
Romaine lettuce (*Lactuca Sativa L.*)	Gas flush with 0, 1, 2.5, 10 or 21 kPa O_2	The samples were sliced, cut, washed in 100 mg/mL chlorine solution (NaOCl)	1. PP with OTR: 8 pmol/s m² Pa 2. PP with OTR: 16.6 pmol/s m² Pa	5 °C—14 days	Severe discoloration was evident on all 16.6 OTR-packaged samples. High O_2 levels resulted in significant alterations of color			Ethanol accumulation in 16.6 OTR packages was less than half of that in 8.0 OTR flushed with ≤10 kPaO₂. Off odors were evident on 8.0 OTR-packaged lettuce flushed with ≤10 kPa O_2	A longer shelf life was achieved from 8.0 OTR-packaged lettuce flushed with mixtures with ≥2.5 kPa O_2	Kim et al. (2005)
Romaine lettuce cv. Paris Island	PMAP		1. LDPE with OTR: 5,676 mL O_2/m² 24 h bar at 20 °C and a 2. Medium density PE with OTR: 4,670 mL O_2/m² 24 h bar at 20 °C 3. PVC with OTR: 3,974 mL O_2/m² 24 h bar at 20 °C	0 and 5 °C—14 days and 18 °C for 12 h	Significant differences in hue angle were detected with samples stored at 0 °C having the best results		Weight loss was reduced on samples stored at 0 and 5 °C (≤1 %). Firmness did not present significant changes with all treated samples differing significantly from the control		Overall quality was better preserved in LDPE and medium density PE bags stored at 0 °C with samples remaining acceptable for the whole storage period	Manolopoulou et al. (2010)

Product	Atmosphere	Packaging	Treatment	Storage	Darkening		Weight loss	Sensory	Visual quality	Reference
Butterhead lettuce (*Lactuca sativa* L.) cv. Wang	Passive MAP: 1. 14 % O_2/5 % CO_2; 2. 16 % O_2/1.2 % CO_2	1. BOPP with OTR: 2,000–3,000 and CDTR: 6,000–7,000; 2. PO PD-961 with OTR: 6,000–8,000 and CDTR: 19,000–22,000 mL/m² day atm.	Treated with chlorinated water (200 ppm total chlorine) for 10 min. Samples were either cut with a sharp knife or by hand	5±0.5 °C—17 days	Dark and necrotic spots were more evident on samples cut with a knife and packaged in BOPP bags		Weight loss increased during storage and reached 5.5 % at the end of storage. A significant decrease in maximum force with storage time was also depicted on the results	Samples in PO bags that were cut by hand had better sensory attributes (later onset of browning of the midribs)	Lettuce hand cut and stored in PO bags showed lower deterioration rates	Martinez et al. (2008)
Winter harvested iceberg lettuce (*Lactuca sativa* L.) cv. Coolguard	1. AMAP with N_2 flush and 5 % O_2 and 0 % CO_2; 2. PMAP	1. Perforated PP (22 μm); 2. PP (25, 30, 40 μm)	All the treatments were vacuum cooled	2 °C—14 days and 12 °C for 2.5 days (shelf-life period)	Brown stains were only found on samples stored in unperforated bags		Weight loss was limited and exceeded 4.75 % only on control samples	Vacuum cooling reduced both the incidence of pink ribs and heart-leaf injury with the effect being evident on the last storage period (last 2.5 days)	Samples stored in either 40 μm PP bags and under AMAP and in 30 μm bags under PMAP had the best overall visual quality	Martinez and Artes (1999)
Lettuce	0.5–2 kPa O_2 balanced with N_2	Film with OTR: 504 mL O_2 and CDTR: 2,507 mL CO_2 25 μm/m² day atm.	Exposure to different light conditions: light (24 h), darkness (24 h) and photoperiod (12 h light and 12 h darkness)	4 °C—3 days and 7 °C—for the rest days	Lightness during storage promoted onset of tissue browning due to increased headspace pO_2			Samples stored under conditions that led to steady state atmospheres of low pO_2 had less off odors and off flavors	Atmosphere conditions with pO_2 being between 0.2 and 0.5 kPa proved to be beneficial for preserving quality of lettuce. Storage under dark had minimal effects on samples quality	Martinez-Sanchez et al. (2011)

(continued)

Table 1.2 (continued)

Species and food type	Initial gas mix	Packaging material	Treatment before packaging	Storage temperature (°C) and storage period (days)	Color	Microflora	Texture-weight loss	Sensory analysis	Shelf life(days)-Life extension	References
Romaine lettuce (*Lactuca sativa* var. Longifolia)	PMAP	1.35 µm PP with OTR: 3,500 mL/m² day atm. 2.35 µm PP film with OTR: 1,100 mL/ m² day atm.	Inoculation with E. coli O157:H7, L. monocytogenes and Salmonella choleraesuis	5 °C—10 days or 25 °C—3 days		A 1 log decrease was noted on E. coli and Salmonella populations while listeria populations were increased by 1 log on samples stored at 5 °C. At 25 °C pathogens increase was between 2.44 and 4.19 log CFU/g. Mesophiles reached 7.8 and 7.5 log CFU/g on samples stored in film 1 and 2 bags, respectively at 5 °C			Pathogen growth on all studied conditions poses a threat and stresses the necessity for protective measures to avoid contamination of vegetables	Oliveira et al. (2010)
Iceberg lettuce	Passive MAP	1. OPP film with OTR: 2,000 mL/m² day atm. 2. PO (RD-106) with OTR: 8,500 mL/m² day atm. 3. PVC film with OTR: 18,500 mL/m² day atm.	Rinsed with tap water (0.2 mg/L total chlorine) for 4 min. Dipping in 0.3 % ascorbic acid and 0.3 % citric acid solution	4 °C for 8 days	Browning was severe on samples from both PO and PVC bags	Mesophile and psychrophile growth was not affected by any treatment or storage condition (6.78 and 6.88 log CFU/g for mesophiles and psychrophiles on the eighth day, respectively)		OPP-packed samples retained better visual quality compared to samples stored in other packaging films	OPP-packed lettuce had a shelf life that exceeded 8 days whereas PVC and PO-packed samples' shelf life was limited to 6 days	Pirovani et al. (1997)

Iceberg lettuce	AMAP: 2 % O_2–98 % N_2	1. Film with OTR: 110 mL O_2/100 in²/m² day atm.	Inoculation of *E. coli* O157:H7 strains	4 and 15 °C for 10 days	Samples stored in bags made from the second film had the lowest *E. coli* populations (3.89 and 6.74 log CFU/g when stored at 4 and 10 °C, respectively)	Even though film 2-packed samples had the lowest microbial load, gene expression analysis showed that these storage conditions supported higher levels of virulence factor expression, while package in film A (typical commercial conditions) had lower virulence expression	Sharma et al. (2011)
		2. The same film with 40 microperforations made with a 25 gauge syringe needle.					
		3. Gas impermeable film					

Francis and O'Beirne (1997) studied the effect of temperature, antimicrobial dips (100 ppm chlorine solution or 1 % citric acid solution for 5 min) and gas atmosphere (initial gas flush with 100 % N_2 or passive MAP) on *L. innocua* and *L. monocytogenes* inoculated on iceberg lettuce. Nitrogen flushing favored the *Listeria* growth more than in unflushed packages while the use of antimicrobial dips led to better survival of *L. innocua* at 3 °C.

Francis and O'Beirne (2001) investigated shredded iceberg lettuce during storage at 4 and 8 °C and under PMAP in regard to the survival and growth of inoculated *L. monocytogenes* and *E. coli O157:H7* [oriented PP (OPP) film used]. *L. monocytogenes* populations rose by approximately 1.5 log cycles during the 12-day storage period at 8 °C, whereas final population densities of *E. coli O157:H7* varied in the range 6.5–7.0 log CFU/g.

Gleeson and O'Beirne (2005) analyzed the effects of different slicing methods during storage (8 °C) on subsequent growth and survival of inoculated *L. innocua* and *E. coli* on passive MAP (PMAP) preserved vegetables (sliced iceberg and butterhead lettuce). The slicing instrument is of great importance since slicing with a blunt knife resulted in consistently higher *E. coli* and *L. innocua* counts during storage than slicing with a razor blade (1.2 log CFU/g lower for *L. innocua* and 0.6 log CFU/g for *E. coli* on butterhead lettuce).

Gomez-Lopez et al (2005) focused on the impact of intense light pulses decontamination on the shelf-life of minimally processed lettuce stored at 7 °C in equilibrium MAP (films with OTR: 2,290 mL/kg h). Five days of experimentation made evident that the psychrotrophic count of treated samples was kept lower (0.46 log CFU/g reduction) than that for the controls whilst yeasts counts amount to 1.8 log CFU/g at day 5 for treated lettuce were higher, but still at low levels.

Iceberg lettuce was treated with gaseous chlorine oxide and cysteine (0.5 % solution) and stored under MAP at 7 °C (initial atmosphere of 2–4 % O_2 and 9 % CO_2) conditions. Microorganisms (mesophiles and psychrotrophs) surviving decontamination grew more rapidly than those present in non-decontaminated samples, as reported by Gomez-Lopez et al. (2008). Therefore, decontamination can considerably accelerate the spoilage rate.

Fresh-cut lettuce, washed with chlorinated water and stored under PMAP (polyolefin laminated film used) was irradiated at a mean dosage of 0.19 kGy. The resulting product had, 8 days after irradiation, microbial population of 290 CFU/g and yeast population of 60 CFU/g, compared with values of 220,000 and 1,400 CFU/g, respectively, for the non-irradiated control. Lettuce irradiated at 0.81 kGy required a mean force of 1,236 N compared to 1,311 N for control (Hagenmaier and Baker 1997).

Jacxsens et al (2002a) used active atmosphere modification (AAM) (3 % O_2 and 5 % CO_2) in order to assess temperature dependence of shelf life of mixed lettuce (mixture of 20 % endive, 20 % curled endive, 20 % radicchio lettuce, 20 % *lollo rosso* and 20 % *lollo bionta* lettuces) as affected by microbial proliferation. The result of the temperature rise was a decrease of the lag-phase of spoilage microorganisms (105 h reduction from 2 to 4 °C). Lettuce retained its organoleptic characteristics best at a storage temperature of 4 °C.

Jacxsens et al (2002b) also investigated packaged minimally processed mixed lettuce under various temperatures (from 5 to 20 °C) in a simulated cold distribution chain, typical of commercial practice, on the microbial quality of EMA (3 % O_2 and 5 % CO_2). The internal atmosphere in the packages remained in its aerobic range during storage in the chain. However, yeasts were shown to be the shelf life limiting group (day 4). Finally, color limited sensorial shelf life down to 5 days.

Jacxsens et al. (2003) attempted to evaluate the quality of mixed lettuce (mixture of endive, curled endive, radicchio lettuce, *lollo rosso* and *lollo bionta* lettuces) with an initial atmosphere of 3 % O_2 and 2–5 % CO_2 and two types of films (low and high permeability). Lettuce stored under EMA (equilibrium modified atmosphere) (high permeability film) had a shelf life of 6 days (limited by growth of psychrotrophs and yeasts) whereas anaerobic conditions in packages of low permeability film resulted in limited shelf life up to 3 days (due to ethanol production release).

Salmonella enteritidis and *L. monocytogenes* was inoculated to fresh lettuce that was afterwards stored under MAP with initial head-spaces of 4.9 % CO_2/2.1 % O_2/93 % N_2 and 5 % CO_2/5.2 % O_2/89.8 % N_2, respectively. Total aerobic counts and lactic acid bacteria (LAB) were lower (0.4 and 0.5 log CFU/g, respectively) in the treated samples. However, both studied pathogens survived but showed no growth regardless of the packaging system applied (Kakiomenou et al. 1998).

Romaine lettuce leaves were sliced, washed, dried and packaged in films with oxygen transmission rates (OTR) of 8.0 and 16.6 pmol s^{-1} m^{-2} Pa^{-1}, and with initial O_2 headspace of 0, 1, 2.5, 10 and 21 kPa and stored at 5 °C. Ethanol accumulation in 16.6 OTR packages was less than half of that in 8.0 OTR packages flushed with ≤10 kPa O_2. For both package types, the onset and the intensity of off-odors was further delayed when flushed with 21 kPa O_2 than flushing with <10 kPa O_2. Samples flushed with 0 or 1 kPa O_2 displayed higher electrolyte leakage than those flushed with higher concentrations of O_2 (Kim et al. 2005).

Romaine lettuce was stored under passive MAP (an LDPE and a medium density PE film was used) under 0 or 5 °C for a storage period of 14 days. Ascorbic acid was better preserved in LDPE packages at 0 °C (81.1 and 77.5 mg/100 g at 0 and 5 °C, respectively). The best visual quality was recorded for samples stored in LDPE bags at 0 °C (Manolopoulou et al. 2010).

The impact of cutting method and packaging film [BOPP and a polyolefin (PO)] on sensory quality of butterhead lettuce stored at 5 °C was studied by Martinez et al. (2008). In comparison with the fresh sample, dark stains significantly developed on the 8th storage day for the lettuce stored in the BOPP film and cut manually with a knife, on the 10th day for the lettuce stored both in the BOPP and PD-961 films when cut, and on the 17th day in the PD-961 film and cut manually.

Winter harvested iceberg lettuce was vacuum-cooled and stored under PMAP (perforated PP, PP 25, 30 and 40 μm) and AMAP (5 % O_2 and 0 % CO_2), while brown stain and heart leaf injury were only detected on heads packaged in unperforated bags. The best treatments for ensuring visual quality were the passive MAP in 40 mm PP and A MAP in 30 mm PP. Pink rib and heart-leaf injury were both reduced by vacuum cooling but the effect only showed up during their shelf-life period (Martinez and Artes 1999).

Romaine lettuce was stored either under AMAP or under PMAP (a film with 507 mL 25 μm/m^2 day atm. was used) and at dark, light and a photoperiod consisting of 12 h of light and 12 h of darkness for 3 days at 4 °C plus 7 days at 7 °C. A steady-state atmosphere of low pO$_2$ (0.2–0.5) led to better overall quality, prevention of browning, off-colors and off-odors. All sensory attributes were negatively affected by light exposure compared to darkness, mainly because of excessive browning effect, due to increased headspace pO$_2$ (Martinez-Sanchez et al. 2011).

Inoculation of romaine lettuce samples with strains of *E. coli O157:H7, L. monocytogenes* and *Salmonella choleraesuis* and subsequently stored under passive MAP (PMAP) [films with 3,500 (film 1) and 1,100 (film 2) mL/m^2 day atm. OTR were used] at 5 and 25 °C for 10 and 3 days, respectively. The storage at 5 °C led to a 1 log decrease of both *Salmonella* and *E. coli* populations while *Listeria* increased by 1 log. The increase of pathogens at higher temperatures varied between 2.44 and 4.19 log CFU/g (Oliveira et al. 2010).

MAs passively developed in the selected packages [mono-oriented PP film (MOPP), PE trays overwrapped with a multilayer PO or PVC film] and dipping in an ascorbic acid (0.3 %) and citric acid (0.3 %) solution at the storage temperature of 4 °C, were studied by Pirovani et al (1998). No substantial changes in psychrotrophic and mesophilic aerobic microbial populations were reported. Lettuce stored in OPP retained better sensory characteristics than products stored in other films.

Inoculation of *E. coli* O157:H7 strains and subsequent storage under MAP A (2 % O$_2$–98 % N$_2$ and a film with OTR 110 mL/m^2 cm^3 day), MAP B (the same film with 40 microperforations) at 4 and 15 °C (for the lettuce inoculated with 5.58 and 3.98 log CFU/g *E. coli O157:H7* respectively) for 10 days resulted in decrease of *E. coli* populations on lettuce at 4 °C under all treatments, but especially under MAP B. Temperature elevation led to a considerable increase in *E. coli* numbers with MAP B displaying the smallest increase (2.5 log CFU/g) (Sharma et al. 2011).

Low storage atmospheres tend to affect *L. monocytogenes* with a descending growth rate as the temperature reduced. *Listeria* strains are prohibited from growing by low oxygen and high carbon dioxide conditions. As *L. innocua* and *L. monocytogenes* have the same growth patterns, their co-existence may have affected both their growth rate negatively (Fig. 1.3).

1.7.3.2 Rocket

Rocket leaves were stored either alone or with the addition of lettuce '*Lollo verde*' leaves under MAP (5 % O$_2$/10 % CO$_2$ for MAP A and 2 % O$_2$/5 % CO$_2$ for MAP B) at 5 ± 1 °C for 10 days. Mesophiles were significantly reduced by both atmosphere modifications 0.7 to 1 log CFU/g reduction, while psychrotrophs were also affected by the initial CO$_2$ (0.7 log CFU reduction at the end of storage). The use of MAP A led to a shelf life extension of 4 days for mixed rocket and lettuce samples remaining acceptable for the whole 10-day period (Arvanitoyannis et al. 2011a, b).

Addition of "*Aceto balsamico di Modena*" wine vinegar and extra virgin olive oil on a mixed lettuce and rocket salad and subsequent storage under MAP

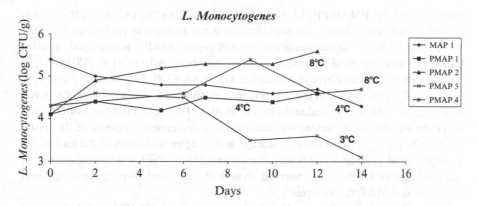

Fig. 1.3 Changes in inoculated *L. monocytogenes* numbers for iceberg lettuce under MAP [MAP 1 (2.1 % O$_2$/4.9 % CO$_2$) with inoculation with *L. monocytogenes*, Kakiomenou et al. (1998), PMAP 1 (with steady-state atmosphere of 3–4 % O$_2$/10–12 % CO$_2$), for iceberg lettuce, PMAP 2 (with steady-state atmosphere of 3–4 % O$_2$/10–12 % CO$_2$),, Francis and O'Beirne (2001), PMAP 3 (with OPP for packaging film) and inoculation with *L. innocua* strains, PMAP 4 (with OPP for packaging film) and inoculation with *L. innocua* strains, Francis and O'Beirne (1997)]. *L. monocytogenes* is affected by low storage atmospheres with a descending growth rate as the temperature is reduced

(5 % O$_2$/10 % CO$_2$ for MAP A and 2 % O$_2$/5 % CO$_2$ for MAP B) at 5 ± 1 °C for 10 days was studied by Arvanitoyannis et al. (2011a, b). The addition of vinegar and both applied MA had a negative effect on microbial growth. Physical attributes were severely affected by vinegar, limiting shelf life to 3 days, while samples with olive oil under MAP A were acceptable for the whole the storage period. Psychrotrophs and mesophiles were reduced in MAP packages with olive oil (1 and 0.5 log CFU/g reductions, respectively) on day 10.

The effect of cutting (whole leaves, cutting into two or four similar parts) combined with passive MAP storage (a film with OTR: 583 mL/m^2 day atm. was used) at 8 °C for 7 or 14 days was investigated by Koukounaras et al. (2010). There were no significant differences in ethylene concentration (0.39 on the 7th day and 0.06 μL/L on the 14th day of storage for the whole leaves) between the different storage conditions. The degree of cutting affected the DPPH radical scavenging activity (16.03, 16.33 and 18.09 Ascorbic acid Equivalent Antioxidant Capacity/100 g FW for whole leaves, cut in half and in four pieces respectively, at the 14th day of storage).

1.7.3.3 Spinach

The effect of super atmospheric O$_2$ (80 and 100 kPa O$_2$ gas flush with two PE films used) and passive MAP (PE$_1$ with high permeability and PE$_2$ barrier films used) on quality of minimally processed baby spinach stored at 5 °C was investigated.

Samples in both PMAP-PE$_2$ (7.7 log CFU/g) and 100-PE$_2$ (8.3 log CFU/g) displayed reductions in aerobic mesophiles compared to those in perforated film (8.7 and 8.8 log CFU/g). Superatmospheric O$_2$ and passive MAP also managed to maintain at low amounts total *Enterobacteriaceae* numbers (Allende et al. 2004a).

The effect of atmosphere modification (12 % O$_2$+7 % CO$_2$ on day 3 and 6 % O$_2$+14 % CO$_2$ on day 7) on the antioxidant constituents of fresh cut spinach stored at 10 °C for 7 days was evaluated by Gil et al. (1999). The effect of both air and MAP on ascorbic acid content was identical to a decrease to one-half of the initial value (630 mg/kg) after 3 days of storage and a higher reduction at the end of the storage period (120 and 100 mg/kg for air and MAP). Within the same period an accumulation of DHAA was observed on both storage conditions (270 and 650 mg/ kg for air and MAP, respectively).

The effect of irradiation on the elimination of pathogens (*Salmonella* spp. and *Listeria* spp.) combined with MAP (100 % O$_2$ and an 1:1 O$_2$:N$_2$ atmosphere) inoculated on baby spinach leaves was investigated. A 5 log reduction on both studied pathogens can be reached by the application of 0.7 kGy irradiation and subsequent storage under superatmospheric MAP. The enhanced O$_2$ in the packages led to a higher irradiation sensitivity (7 to 25 % reduction in D$_{10}$ values) that can be attributed to the produced ozone.

Kaur et al. (2011) investigated the effect of MAP storage (LDPE and PP films were used) on spinach leaves of variable in-pack weight (200, 400 and 600 g of leaves per pack) stored at 15 °C for 4 days. Carotene and ascorbic acid were better preserved in LDPE packages (22 to 24 and 16 to 23 mg/100 g for carotene in LDPE packages and PP packages, respectively) compared to PP ones. Phenolic compounds were higher in LDPE packages; 190 to 250 and 30 to 80 mg/100 g FW in LDPE packages and PP packages, respectively.

According to Lee and Baek (2008), spinach inoculated with *E. coli O157:H7* was packed under four different environments (air, vacuum, 100 % N$_2$ gas, and 100 % CO$_2$ gas packaging) following treatment with water, 100 ppm chlorine dioxide, or 100 ppm sodium hypochlorite for 5 min and stored at 7±2 °C. The lowest levels of *E. coli O157:H7* were detected in samples ClO$_2$ treated and stored under vacuum; 1.8, 2.3, 2.6 log CFU/g and 3.9, 3.1, and 4.0 log CFU/g for samples packaged in vacuum, N$_2$ gas, and CO$_2$ gas following treatment with ClO$_2$ and NaOCl, respectively after 7 days of storage.

Fresh-cut spinach was treated with citric acid and ascorbic acid solutions (variations from 0 % to 1 %) and packaged in OPP or LDPE bags (passive MAP) and stored at 4 °C. Mesophiles were affected and their number decreased as citric acid concentration rose. Even though the citric and ascorbic acid treatment reduced the pH and the initial microbial load, these effects were neutralized over storage (Piagentini et al. 2003).

Rodriguez-Hidalgo et al. (2010) examined the effect of fertilization (8, 12 or 16 mmol N L^{-1}) combined with MAP storage (passive with a BOPP film used, a N$_2$O enriched atmosphere and a superatmospheric atmosphere with 100 % O$_2$) on quality of fresh baby spinach leaves stored at 5 °C for 10 days. The use of fertilizers (12 and 16 mmol concentrations) and N$_2$O MAP led to the lowest microbial growth

(6.7 and 6.4 log CFU/g for mesophiles for 12 and 16 mmol concentrations, respectively) with overall good sensory quality after 8 days of storage. Total antioxidant capacity of these samples was preserved close to the initial levels (8 g ascorbic acid equivalent antioxidant capacity kg^{-1} FW).

1.7.3.4 Cabbage

Cut salted Chinese cabbage with air, 100 % CO_2 or 25 % CO_2/75 % N_2 packaging was irradiated with 0.5, 1 and 2 kGy and the microbiological and physicochemical parameters were investigated during storage at 4 °C for 3 weeks. Gamma irradiation at 0.5 kGy was found to have a significant effect on initial TVC (2–3 log CFU/g reduction) that continued with MAP storage (0.9–1.9 log CFU/g reduction compared to aerobic packaging). Texture was maintained close to the values of the fresh product on MAP samples (Ahn et al. 2005).

Shredded cabbage was immersed in neutral electrolyzed oxidizing water (NEW) containing 40 mg/L of free chlorine up to 5 min, and then stored under EMAP (film with 4,600 mL O_2/m^2 24 h atm. was used at 7 °C) at 4 and 7 °C. Water washed samples had higher pH values than those undergone NEW treatment (6.17 and 6.07, respectively). LAB counts varied from 1.9 up to 3.5 log CFU/g throughout the study. Overall visual quality (OVQ) was the key factor in determining the shredded cabbage shelf life. Treated samples showed a shelf life extension of 3 days (shelf life of control samples was 6 days) and were finally rejected due to poor OVQ, browning and dryness (Gomez-Lopez et al. 2007b).

The combination of AMAP (MAP A: 70 kPa O_2 and 15 kPa CO_2 and MAP B: 5 kPa O_2 and 15 kPa CO_2) and MVP with different type of films (Ny/PE with OTR: 54.8±0.7 mL/m² day atm. and LDPE with OTR: 1.277±159 mL/m² day atm.) was used for fresh cabbage preservation for 10 days at 5 °C. The use of MAP A with both films led to a significant microbial inhibition (4.82, 4.28, 4.99, 5.2, 4.28 and 4.91 log CFU/g for *P. fluorescens*, *E. coli*, *E. coli O157:H7*, *S. Typhimurium*, *S. aureus* and *L. monocytogenes*, respectively for MAP A and PE). Application of barrier films led to better appearance scores (Lee et al. 2011).

Pirovani et al. (1997) investigated the effect of passive MAP [OPP (treat. 1), PE trays overwrapped with multilayer polyolefin (PO) (treat. 2), or with a plasticized PVC (treat. 3)] and storage at 3 °C on quality of cabbage. The weight loss remained within acceptable levels for all 3 treatments (0.08, 0.4 and 0.93 % for treat. 1, 2 and 3 respectively). General appearance of samples in OPP bags was better preserved exhibiting less wilting and browning.

1.7.3.5 Kale

Kobori et al (2011) investigated the effect of different storage temperatures (1 and 11 °C) and storage conditions (light exposure) on the shelf life and behavior of flavonols and carotenoids of kale leaves stored under passive MAP (stretched PVC film with OTR: 12,889 mL/m^2 day). Shelf life was limited to 6 and 3 days for samples stored at 11 °C at dark and light conditions respectively while kale leaves at 1 °C had their sensory quality decreased during the 17-day storage period. Quercetin and kaempferol were found to be relatively stable at light exposure positively affecting them. Neoxanthin, violaxanthin, lutein, and β-carotene decreased (16.1, 13.2, 24.1 and 23.7 %, respectively) after 10 days at 11 °C in the dark.

1.7.3.6 Betel Leaves

Rai et al. (2010) investigated the effect of atmosphere modification (a PP film was used with OTR: 1.49×10^{-6} mL m/m^2/h/kPa and with three different in-pack weights 250, 500 and 750 g) on betel leaves stored at 20 °C for 10 days. Chlorophyll was better preserved in 750 g packages. Presence of high CO_2 atmospheres proved to be capable of preserving chlorophyll and preventing browning of the leaves.

1.7.3.7 Chicory Endive

Airtight containers with continuous flow of gas mixture (1.5 % O_2 and 20 % CO_2/0 % O_2 and 20 % CO_2) were used by Bennik et al (1996) to monitor the growth of inoculated *L. monocytogenes* on chicory endives stored at 8 °C. The rate of spoilage proceeded much rapidly at ambient conditions, since all subpopulations reached the 10^6 CFU/g level within 2 days of storage. Four days were adequate for this level to be reached under low O_2 and high CO_2 contents.

Minimally processed fresh broad-leaved endives were stored at 3 and 10 °C in modified atmospheres containing 10 % O_2/10 % CO_2, 10 % O_2/30 % CO_2 and 10 % O_2/50 % CO_2 and control samples (air). The highest increase in aerobic bacteria occurred in air (78 %), then in 10 % O_2/10 % CO_2 (67 %), whereas the lowest increase at 10 % O_2/50 % CO_2 (51 %). At 10 °C inoculated *L. monocytogenes* tended to grow very rapidly as the concentration of CO_2 increased (8–8.2 log CFU/g at 50 % CO_2) as proved by Carlin et al. (1996).

Charles et al. (2008) utilized oxygen scavengers to make up for modified atmosphere conditions at LDPE bags to investigate the effect of packaging conditions on quality changes of endives stored at 20 °C. A higher greening in control [macroperforated oriented polypropylene (MPOPP) and passive MAP (LDPE pouches without oxygen scavengers)] samples than in A MAP was evident (3, 2.7 and 2 for control, passive and active MAP samples, respectively after 7 days of storage).

High oxygen atmosphere (70, 80 and 95 % O_2) was also used by Jacxsens et al. (2001) as an alternative for EMA to prolong the shelf life of shredded chicory endive. The packages of shredded chicory endives under low O_2 concentrations exceeded the limit for yeasts (10^5 CFU/g) by the fourth day while the high O_2 packages exceeded the limit by the seventh day. The evolution of the enzymatic discoloration of chicory endive evolved rapidly under EMA limiting shelf life to 3 days, whereas no unacceptable scores were obtained for the MAP packaged vegetables.

Jacxsens et al. (2003) evaluated the quality of shredded chicory endives, packaged in consumer-sized packages under EMA (initial atmosphere of 3 % O_2 and 2–5 % CO_2) and stored at 7 °C. Shredded chicory endives were found to have after day 8 more than 10^8 CFU/g total psychrotrophic counts. The pH of the product remained stable around 6 over the entire storage period of 13 days. The color was shown to be the limiting sensory property at day 6.

The effect of irradiation (0.3 or 0.6 kGy) combined with passive (laminated foil/ plastic barrier bag) or A MAP (5 % O_2 and 5 % CO_2/10 % O_2 and 10 % CO_2) on survival of inoculated *L. monocytogenes* and sensory qualities of endive stored at 4 °C was investigated by Niemira et al. (2005). *L. monocytogenes* and total microbial populations on the irradiated MAP samples were either lower than or not different compared to the initial levels, whereas irradiated leaf material in air retained their color attributes over storage better than non-irradiated.

1.7.4   Fruits–Vegetables

1.7.4.1   Tomato

Active (3 kPa O_2+0 kPa CO_2 and 3 kPa O_2+4 kPa CO_2) and passive (OPP film of 35 μm thickness) MAPs were used at 0 and 5 °C for tomato storage. The highest C_2H_4 level was found in passive MAP packages at 5 °C (15 μL/L) while in A MAP, at both temperatures about 6 μL/L C_2H_4 were accumulated. Sensorial parameters were preserved by both passive and AMAP above the acceptable limit of marketability and lowered the weight losses even at higher temperatures (5 °C) (Aguayo et al. 2004).

Artes et al. (1999) investigated the effect of calcium chloride washings (0.7 mM chlorinated water with or without 0.09 M $CaCl_2$) in conjunction with passive or AMAP (7.5 % O_2) stored at 2 and 10 °C on fresh-cut tomato quality preservation. In samples that were kept at 10 °C, a significant reduction in firmness (38.5 and 37.3 % for passive and A MAP) was recorded. At 10 °C, tomato slices under passive or AMAP did not reveal any change in color compared with the other treatments.

Bailen et al. (2006) assessed the use of granular-activated carbon (GAC) alone or impregnated with palladium as a catalyst inside tomato packages (OPP film used) under PMAP. Ethylene was lower in MAP packages with GAC and especially in those with GAC-Pd (8 μL/L). Control tomatoes showed the highest weight loss (0.72 %), although weight loss was very low. No off-flavor was detected in those packages with GAC and especially with GAC-Pd.

Four different packaging films (PE-20 μ and 50 μ, PVC-10 μ, PP-25 μ) were used by Batu and Thompson (1998) to provide the ideal conditions to preserve pink tomatoes stored at 13 °C. While in storage, all fruit softened progressively but those sealed in plastic films softened significantly more slowly than those stored unwrapped whereas PP and PE_{50} had both the lowest gas permeability and weight loss (1.6–1.9 %).

The effect of passive MAP (LDPE film used), CA (5 % CO_2) and gaseous ozone treatment in regard with the survival of inoculated (low and high inoculum levels) *S. enteritidis* on cherry tomatoes was investigated by Das et al. (2006). The *S. enteritidis* had a higher death rate on the tomato surfaces stored in MAP than on tomatoes stored in air and in CA. In an initial population of 3.0 $\log_{10}$ CFU/tomato, cells died completely on day 4 during MAP storage and on day 6 during both CA and air storage. Moreover, gaseous ozone treatment was shown to have a strong bactericidal effect on *S. enteritidis*.

Fresh cut tomato slices at 0 and 5 °C under active (12–14 kPa O_2+0 kPa CO_2) MAP [Composite (Vascolan) or BOPP film] with or without an ethylene absorber (EAP) were packaged by Gil et al. (2002). Decrease of titratable acidity was recorded over storage at both temperatures, except for tomato slices kept in composite packages with EAP (0.34 and 0.32 for 0 and 5 °C, respectively) and for control samples packed in composite film at 5 °C (0.33).

Cold storage (5–10 °C) was used to assess the quality of fresh-cut tomato slices under various MAP conditions (4 % CO_2+1 or 20 % O_2, 8 % CO_2+1 or 20 % O_2, or 12 % CO_2+1 or 20 % O_2) and various packaging films [film A and Film B (87.4 and 60 mL/h m^2 atm., respectively)]. Ethylene concentration in containers with four slices held at 5 °C was higher with film B than for film A. The slices in containers of 12 % CO_2+1 % O_2 gave the highest firmness values (3.7 N). Application of MAP resulted in good quality tomato slices with a shelf life of 2 weeks or more at 5 °C (Hong and Gross 2001).

Six different tomato cultivars were stored under MAP (5 % O_2+5 % CO_2) and refrigeration (4 °C) in a study by Odriozola-Serrano et al. (2008). A significant increase in phenolic compounds was recorded beyond the 14th day of storage [maximum values of 347.5 mg gallic acid/kg (FW) for fresh-cut Durinta tomatoes]. Overall, the main antioxidant compounds (lycopene, vitamin C and phenolic compounds) and color parameters were preserved in fresh-cut tomatoes for 21 days.

The effect of 1-methylcyclopropene (1-MCP) (1,000 nL/L dosage) in combination with PMAP (a film with CO_2TR, 2,023 cm^3/m^2 day) on quality of pink or light red tomatoes stored at 12 °C for 21 days was studied by Sabir and Agar (2011). Weight was better preserved with MAP and 1-MCP combination (1.43 % weight loss at the end of the storage period). Lycopene accumulation was preserved under MAP and 1-MCP treatment (7.48, 15.61 and 16.1 mg/kg for treated, MAP and control samples, respectively). Storage life of tomatoes was extended up to 21 days with MAP and 1-MCP application (Table 1.3).

Mature green cherry tomato was dipped in hot water (at 39 °C for 90 min) and subsequently stored in plastic films with various O_2 but similar CO_2 permeabilities at 15 °C. The fruit subjected to HWT with the low O_2 permeability film demonstrated the lowest color development (33.04, 41.46 and 1.13 C*, h° and a*/b* values, respectively). Thus, the heat treatment in conjunction with film packaging suppressed the color development of cherry tomato more than that of individual heat treatment or packaging (Sayed Ali et al. 2004).

Tomatoes were immersed in hot water (42.5 °C) for 30 min and stored under passive MAP (LDPE film) for 14 days at 10 °C and then at 22 °C for 3 days without

Table 1.3 Analysis of the storage conditions, pretreatments and their effect on the storage life of tomato

Species and food type	Initial gas mix	Package material	Treatment prior to packaging	Storage temperature (°C) and storage period (days)	Treatment effect and shelf life extension	References
Cherry tomato (*Lycopersicon esculentum Mill.* cvs. "Alona" and "Cluster")	1. PMAP 1 with 50 µm PE used 2. PMAP 2 with 100 µm PE used	1. 50 µm PE with OTR: 64.27 cm^3 m^2 gun^{-1} and CO$_2$TR: 303.2 cm^3 m^2 gun^{-1} 2. 100 µm PE with OTR: 116.6 cm^3 m^2 gun^{-1} and CO$_2$TR: 612.4 cm^3 m^2 gun^{-1}	Addition of harpin on the pre-harvest period	5–7 °C—28 days	Harpin in conjunction with PMAP 1 were the most effective treatments for preventing maturity of cherry tomatoes	Akbudak et al. (2012)
Sliced roma tomatoes	1. 10/90 % O$_2$/N$_2$ 2. 5/95 % O$_2$/N$_2$ 3. 100 % N$_2$	A Multivac A 300/16 gas packaging system	Immediate irradiation or after 24 or 48 h after storage. Inoculation with *Salmonella* strains	10 °C—28 days	*Salmonella* populations were reduced under all treatments with the least dose necessary being at control samples that were irradiated immediately	Niemira and Boyd (2013)
Tomatoes (*Lycopersicon esculentum,* var. Marglobe)	1. PMAP 2. Air	Microperforated bags using Xtend® film	Comcat solution was applied during the preharvest period. Postharvest dipping into chlorine and NaOCl solution	13 °C or room temperature—30 days	The combination of Comcat pretreatment and subsequent storage at 13 °C under MAP was the most effective conditions for retaining tomato quality parameters (15.7 g/100 g, 4.6, 3.1 and 5.8 log CFU/g for ascorbic acid, TVC, fungi and coliforms after 30 days of storage)	Workneh et al. (2011)

(continued)

Table 1.3 (continued)

Species and food type	Initial gas mix	Package material	Treatment prior to packaging	Storage temperature (°C) and storage period (days)	Treatment effect and shelf life extension	References
Tomatoes (*Lycopersicon esculentum* Mill.)	1.5 %/5 %/90 % $O_2/CO_2/N_2$	32 μm thick commercial plastic bag (FRESHPAC) with OTR: 11.66 mL/m² day and CO_2TR: 27.97 mL/m² day	Tomatoes were vaporized with 10^{-4} M of methyl jasmonate solution	5 °C—63 days	Methyl jasmonate treated tomatoes stored under MAP showed a shelf life extension of 3 weeks (9 weeks of shelf life) compared to control samples (109.36 N firmness, 1.42 % weight loss, 0.23 % citric acid, after 63 days of storage)	Siripatrawan and Assatarakul (2009)
Vine-ripened tomato fruits (*Solanum lycopersicum* L.)	1. PMAP 1	1. Three-layer laminated (PET/ aluminum/polyethylene) high barrier pouches were used to create high CO2 conditions.	Packages contained 10 g of C_2H_4 absorber	15 °C—12.6 days and 25 °C—6.6 days	GABA concentration increased on the second PMAP by 75 % compared to control samples as storage temperature increased	Mae et al. (2010)
	2. PMAP 2	2. Micro-perforated PET/ LDPE film pouches				
Tomato var. superjeff	1. CA 5 kPa/3 kPa O_2/CO_2	For MAP a PE bag was used		13 °C—90 days	The application of MAP and CAS managed to delay the ripening processes of tomatoes (5.18 °Brix for total soluble solids for CAS on the 90th day and 0.439 and 0.481 titratable acidity for MAP and CAS on the 40th day). CAS managed to prolong shelf life at 90 days better than MAP and cold storage	Majidi et al. (2011)
	2. MAP 5 kPa/3 kPa O_2/CO_2					

Product	Treatment	Packaging details	Dipping treatment	Storage	Results	Reference
Cherry tomato (*Lycopersicon esculentum*) cv. Coco	1. PMAP 1 with M1 film 2. PMAP 2 with M2 film 3. PMAP 3 with M3 film	Three PE packages were used: 1. M1 with OTR: 7.78 and CO_2TR: 9.12×10^{-17} mol s^{-1} m^{-2} Pa^{-1} 2. M2 with OTR: 5.49 and CO_2TR: 8.47×10^{-17} mol s^{-1} m^{-2} Pa^{-1} 3. M3 with OTR: 3.56×10^{-17} mol s^{-1} m^{-2} Pa^{-1} and CO_2TR: 8.97×10^{-17} mol s^{-1} m^{-2} Pa^{-1}	Dipping in hot water (39 °C for 90 min)	15 °C—13 days	Hot water treatment combined with low O_2 atmosphere packaging delayed color development on the studied samples. Acceptability was prolonged by 2, 4 and 6 days on samples stored under PMAP 1, PMAP 2 and PMAP 3 respectively, compared to the control samples	Sayed Ali et al. (2004)
"Maru" tomatoes (*Lycopersicon esculentum* Mill.)	PMAP	LDPE film was used	Dipping in hot water (42.5 °C for 30 min)	10 °C—14 days and 3 days at 20 °C without packaging	The use of hot water dipping and subsequent MAP storage reduced weight loss (1.74 %) and unwanted changes of color and firmness whereas soluble solids (6.47 °Brix) and titratable acidity (0.63 %) remained unaffected	Suparlan and Itoh (2003)
Tomato (*Lycopersicum esculentum* Mill.) cvrs Rambo, Durinta, Bodar, Pitenza, Cencara, and Bola	MAP 5 kPa/5 kPa O_2/CO_2	ILPRA film with OTR: 110 cm³/m² day bar and CO_2TR: 500 cm³/m² day bar	Dipping for 2 min in chlorinated water (0.2 mg free chlorine/L). Samples were sliced (7 mm-thick slices)	4 °C—21 days	Lycopene content peaked on Bodar cultivar (80.5 mg/kg). Durinta tomatoes had the highest phenolic content and vitamin C (314.1 and 204.8 mg/kg). Color attributes and antioxidant content was preserved for the whole storage period of 21 days on fresh cut tomatoes	Odriozola-Serrano et al. (2008)

(continued)

Table 1.3 (continued)

Species and food type	Initial gas mix	Package material	Treatment prior to packaging	Storage temperature (°C) and storage period (days)	Treatment effect and shelf life extension	References
Tomato (cv 'Liberto')	1. PMAP 1 created by the first film 2. PMAP 2 3. PMAP 3 4. PMAP 4	1. PE 20 µm 2. PE 50 µm 3. PVC 10 µm 4. PP 25 µm	Dipping for 1–2 min into 100 ppm of Thiabendazole solution	13 °C—60 days	Samples stored under PMAP 2 and 4 maintained their characteristics until the 60th day of storage (1.5 and 1.6 Nt/mm for firmness, 1.2 and 1 % for weight loss of MAP 2 and 4 samples on the 60th day)	Batu and Thompson (1998)
Tomato (*Lycopersicum esculentum Mill.*) cv Durinta	1. PMAP 2. Active MAP 7.5 % O_2	Vascolan film with OTR: 2.35 and CO_2TR: 6.11×10^{-14} mol s^{-1} m^{-2} Pa^{-1}	Washing with chlorinated water (0.07 mM) with or without $CaCl_2$ (0.09 M)	2 and 10 °C—7 and 10 days	Optimal results were found under MAP storage at 2 °C (58.5 and 52.4 N for firmness, 4.3 and 4.2°Brix for soluble solids, 0.33 and 0.34 g citric acid/100 g for titratable acidity of passive and AMAP samples, respectively on the tenth day). $CaCl_2$ dips improved samples attributes only at low storage atmosphere (2 °C)	Artes et al. (1999)

Sample	Gas composition	Packaging film		Pretreatment	Storage	Results	Reference
Tomato slices (*Lycopersicon esculentum*) cv Mill.	1. 1 %/4 % O₂/CO₂ 2. 20 %/4 % O₂/CO₂ 3. 1 %/8 % O₂/CO₂ 4. 20 %/8 % O₂/CO₂ 5. 1 %/12 % O₂/CO₂ 6. 20 %/12 % O₂/CO₂ 7. Control samples	Film A with OTR: 87.4 at 5 °C and 119.3 mL h⁻¹ m⁻² atm.⁻¹ at 10 °C and CO₂TR: 493.4 mL h⁻¹ m⁻² atm.⁻¹	Film B with OTR: 60 at 5 °C and 77.8 mL h⁻¹ m⁻² atm.⁻¹ at 10 °C and CO₂TR: 210 mL h⁻¹ m⁻² atm.⁻¹		5 or 10 °C for 19 days	Samples stored under 1 %/12 % O₂/CO₂ had the highest firmness values (3.7) and no fungal growth. Storage at low temperature can lead to a shelf life of 2 weeks under MAP storage	Hong and Gross (2001)
Tomato slices (*Lycopersicum esculentum Mill.*) cv. Durinta	12–14 kPa/0 kPa O₂/CO₂ An ethylene absorber was added (KMnO₄ on celite)	1. Vascolan with OTR: 2.4 and CO₂TR: 6.1×10⁻¹⁴ mol s⁻¹ m⁻² Pa⁻¹ 2. BOPP with OTR: 3.3 and CO₂TR: 3,100×10⁻¹² mol s⁻¹ m⁻² Pa⁻¹		Dipping into sodium hypochlorite solution (1.3 mM) for 1 min	0 or 5 °C for 7 and 10 days	High CO₂ atmosphere and storage temperature of 5 °C was the best storage combination for tomatoes (78.8 for Hue angle, 0.33 g citric acid/100 g for titratable acidity and 14.5 for soluble solids to titratable acidity ratio on the tenth day)	Gil et al. (2002)

(continued)

Table 1.3 (continued)

Species and food type	Initial gas mix	Package material	Treatment prior to packaging	Storage temperature (°C) and storage period (days)	Treatment effect and shelf life extension	References
Tomato (*Lycopersicum esculentum* Mill.) cv. Calibra	1. PMAP 2. 3 kPa O_2 3. 3 kPa/4 kPa O_2/CO_2	An OPP film of 35 μm thickness was used with OTR: 5.5 L/m² day atm. and CO_2TR: 10 L/m² day atm.	Dipping into sodium hypochlorite solution (1.3 mM) for 1 min. Samples were sliced or divided into wedges	0 and 5 °C—14 days	A shelf life of 14 days was achieved with the combination of AMAP and storage at 0 °C (5–7 μL/L for ethylene levels, 3.2–3.4 log CFU/g for TVC on tomato slices, <2 log CFU/g for yeasts and molds)	Aguayo et al. (2004)
Tomato (*Lycopersicon esculentum* Mill.) cv. Beef	1. PMAP 1 2. PMAP 2	An OPP film of 35 μm thickness was used with OTR: 1,600 mL/m² day atm. and CO_2TR: 3,600 mL/m² day atm.	Granular-activated carbon (GAC) was induced either alone or impregnated with palladium as a catalyst (GAC-Pd) (ethylene absorbers)		The limit of 40 % on decay incidence was reached on the 21st for the control samples while tomatoes stored with GAC-Pd absorbers stayed acceptable until the 28th day of storage	Bailen et al. (2006)

packaging. The rates of ethylene production of HWT tomatoes were slightly higher than controls (28.4 and 23.1 ppm for treated and untreated samples after 4 days). Suparlan and Itoh (2003) suggested that the HWT and MAP usage caused the reduction of weight loss and decay and the inhibition of color development as well as the preservation of firmness.

Mature green tomatoes stored in MA containers with steady state atmospheres of 5 % O_2 and 5 % CO_2 at 13 °C and subjected to a sequence of temperature fluctuations ($\Delta T = 10$ °C) over 35 days to simulate storage and transport conditions were studied by Tano et al. (2007). Accumulation of higher levels of both ethanol and acetaldehyde was observed in tomatoes stored under MA (49.5 and 5.1 mg/kg, respectively) and that tendency became more apparent with temperature fluctuation conditions.

1.7.4.2 Eggplant

Arvanitoyannis et al. (2005) examined the effect of grafting [with *S. sisymbrifolium* (gr. 1), *S. torvum* (gr. 2), methylbromide (gr. 3) and Perlka (gr. 4)] and storage under MAP (30 % CO_2 in HDPE bags) at 10 °C on quality parameters of eggplant. Vitamin C decreased after 7 days and its drop became substantial after 14 days (27, 34, 33, and 45 mg/100 g for gr. 1, 2, 3 and 4, respectively). Storage under MAP resulted in better maintenance of saltiness, acridness, grassiness, metal, hardness and overall impression, in comparison with the values obtained air storage.

1.7.4.3 Cucumbers

The effect of storage conditions [storage under CA at normal and superatmospheric (70 % O_2) conditions] and physical tissue damage (bruising due to weight drop) on membrane peroxidation in minimally processed cucumber tissue stored at 4 °C or 20 °C was studied. Bruised samples showed a continuous increase in lipid hydroperoxides levels (5.5 and 5.2 μmol H_2O_2 equivalents/g for normal and superatmospheric conditions, respectively) whereas in non-bruised samples there was a different pattern with an increase within the first three days followed by a decrease for the rest of the storage time (Karakas and Yildiz 2007).

Wang and Qi (1997) investigated the cucumbers storage under passive MA conditions (microperforated and intact LDPE films) at 5 °C for 18 days. MAP stored samples showed chilling injury symptoms on the 12th day of storage while control samples had significant losses starting on the 6th day. Samples in perforated bags were found unacceptable because of decay occurrence on the 15th day of storage. Samples in sealed packages did not show any sign of off-flavor even at the end of the experiment.

1.7.4.4   Pepper

Gonzalez-Aguilar et al. (2004) investigated the effect of VP (Saran film used) and passive MAP (PD-961 polyolefin films used) at 5 and 10 °C on quality and shelf life of bell peppers. At 5 and 10 °C the ethanol and acetaldehyde contents were significantly lower (0.04 and 0.18 μL/g, at 5 °C and 0.36 and 0.5 μL/g at 10 °C for ethanol and acetaldehyde, respectively) on peppers under MAP compared to samples under vacuum (0.41 and 0.78 μL/g for ethanol and acetaldehyde, respectively). The shelf life at 5 and 10 °C was limited to 21 and 14 days, respectively due to microbiological and quality parameters.

A combination of biodegradable film (PLA based film) and LDPE (perforated or not) was used to provide MA conditions for storage of green peppers at 10 °C for 7 days. Hue angle and chroma did not display any considerable change in all treatments. Aerobic bacteria populations did not increase significantly (<1 log CFU/g) in PLA, LDPE and p-LDPE film packages and the same pattern was depicted in coliform growth (0.2, 2.3 and 0.9 log CFU/g increase for PLA, LDPE and p-LDPE, respectively) (Koide and Shi 2007).

Green chili peppers were stored under passive MAP (LDPE, PVC and cast PP films were used) at 10 °C for 14 days. Ascorbic acid was higher in packaged peppers compared to control samples (67.4, 69.3, 64.8 and 63.3 mg/100 g for LDPE, CPP, PVC and control samples, respectively). The equilibrium atmospheres in LDPE and cast PP packages were very close to the optimal gas concentrations (3 % O_2 and 5 % CO_2) and so they had a beneficial role in quality maintenance (Lee et al. 1994).

Sliced bell peppers were washed once, twice or three sequential times in fresh distilled water and stored under MAP (polyethylene film used) at 7 °C for 10 days. Packages with unwashed samples always had higher CO_2 content. Washed slices were significantly firmer than unwashed and this effect improved incrementally with the number of washes (284, 287, 288 and 293 g/force for unwashed, 1, 2 and 3 washes, respectively) (Toivonen and Stan 2004).

Wall and Berghage (1996) used pressed cardboard trays overwrapped with VF-71 PE and LDPE bags for MA storage of green chile peppers at 8 or 24 °C. Weight loss at 8 °C was 18 %, 6 % and 0.3 % for control, PE and LDPE storage respectively after 6 weeks while at 24 °C the loss was 68 %, 37 % and 0.6 % for control, PE and LDPE storage, respectively after 4 weeks. The postharvest disease ratings for peppers stored in trays were 1.0 and 1.6 after 2 and 4 weeks at 24 °C respectively, whereas the ratings for peppers in bags were 1.4 and 1.6 for the same time periods.

1.7.4.5   Pumpkin

Cut pieces of pumpkin were dip treated [in citric acid (0.2 %) and potassium metabisulfite (0.1 %) for 3 min (treat. 1), soluble starch (0.2 %) extrapure plus calcium chloride (0.1 %) for 3 min (treat. 2) and mannose (0.1 %), and vacuumized for 5 min (treat 3)] and stored in LDPE or PP bags, sealed both with and without vacuum and stored at 5 ± 2 °C, 13 ± 2 °C and 23 ± 2 °C. All samples stored under 5 ± 2 °C became gradually softer toward the end of storage (LDPE decrease 16.9 %). The shelf life of treated diced pumpkin stored in LDPE film bags at 5 ± 2 °C was extended to 25 days (Habibunnisa et al. 2001).

1.7.4.6 Zucchini

Two different zucchini cultivars (Sofia and Diamante cultivars) were sliced and subsequently stored in bags from two packaging materials [OPP with 90 μm thickness and biodegradable coextruded polyesters (CoEX) with 35 μm thickness] under both PMAP and AMAP (5 % O_2/5 % CO_2/90 % N_2 for MAP 1 with OPP and 15 % O_2/10 % CO_2/75 % N_2 for MAP 2 with CoEX) at 5 °C for 9 days. Zucchini slices stored either passively or under AMAP into OPP bags had longer shelf lives (6 and 7 days for Diamante cultivar and 5 and 3 days for Sofia cultivar under AMAP and PMAP, respectively) (Lucera et al. 2010).

1.7.5 Bulbs

1.7.5.1 Garlic

Fresh garlic sprouts were stored under passive MAP (PVC with OTR: 5,500 mL/ m^2 day atm. and CO_2TR: 10,000 mL/m^2 day atm. and LDPE with OTR: 7,000 mL/ m^2 day atm. and CO_2TR: 35,000 mL/m^2 day atm. films were used) for 15 days at 4 °C. Microbial growth was limited by MAP conditions (a 3 log reduction for psychrotrophs in PVC samples compared to control). PVC packages gave the best results for keeping the overall quality of garlic sprouts (Li et al. 2010).

1.7.5.2 Onions

Five different packaging treatments (PMAP with LDPE and PP, A MAP with LDPE and an ethylene scavenger used, LDPE with initial gas concentration of 9.5 kPa $CO_2 + 18.2$ kPa O_2 and moderate vacuum packaging) were used for bunched onions stored at 10 °C for 28 days to determine the optimum packing method. Samples packaged in PE + ES displayed greater weight loss (3.1 % at 14th day) than for other treatments. MVP limited microbial growth more effectively than the rest of the treatments (Hong and Kim 2004).

Liu and Li (2006) tested the microbial proliferation and sensory quality aspects of sliced onions at three temperatures (−2 °C, 4 °C and 10 °C) and atmospheric conditions [LDPE with (A MAP) or without (PMAP) 40 % $CO_2 + 1$ % O_2]. The b^* values were significantly affected by the storage temperature and were higher at 10 °C than at −2 and 4 °C under the same storage conditions after 17 days of storage. Spoilage of the product was mainly attributed to psychrotroph growth.

1.7.6   Stems and Shoots

1.7.6.1   Kohlrabi

Escalona et al. (2007a) studied the effect of passive MAP (with OPP and amide PE films used after washing in a NaOCl water solution) to preserve kohlrabi sticks at 0 °C for 14 days. Kohlrabi reached a very low C_2H_4 release of 5–10 nL C_2H_4/kg/h from the beginning till the end of storage. The L^*, a^*, b^*, and chroma parameters decreased after 7 days at in all treatments (36.3, 36.1, 36.5, for L^*, −2.5, −2.6, −2.4 for a^*, 2.4, 2.3, 1.8 for b^* and 3.5, 3.5 and 3 for chroma control, OPP and Amide PE samples respectively, after 14 days of storage). The sticks stored under MAP conditions scored above the limit of marketability for sensorial attributes on the 14th day of storage.

Kohlrabi stems were stored under MAP (with OPP and Amide-PE and washed in a NaOCl water solution) for 60 days at 0 °C, while an additional retail sale period of 3 days at 12 °C after each cold storage evaluation (30 and 60 days) was applied. A minor yellowing representing lower hue value on the skin after 60+3 days appeared on the MAP kohlrabi stems. The weight loss on control stems from day 30 to 60, as indicated by Escalona et al. (2007b) for amide-PE and OPP were 0.24, 0.31 and 0.1 %, respectively.

Passive MAP (OPP 20 μm, OPP 40 μm, and amide polyethylene films) applied on stored kohlrabi stems at 0 °C for 14 days and at 10 °C for 3 days was monitored by Escalona et al. (2007c). The ethylene production was higher than 0.05 μL/kg/h throughout storage. The citric acid content (79.3 mg/100 mL at harvest) did not display any significant changes (64.8, 73, 72.2 mg/100 mL for OPP_{20}, OPP_{40} and amide-PE respectively). Although the appearance and texture of kohlrabi deteriorated compared to values at harvest, the scores were maintained throughout the retail sale period.

1.7.6.2   Bamboo Shoots

An effort was made to prolong the shelf life of bamboo shoots by applying MAP (LDPE film with initial atmosphere of 2 % O_2 and 5 % CO_2) on bamboo shoots stored at 10 °C by Shen and his co-workers (2006). A browning process began in the bamboo shoots of MAP treatment at the sixth day and their browning index reached 1.2 on day 10. MAP treatment inhibited the activity of POD (32 with 55 u/min g FW for the control samples) and PAL (1.25 with 2.0 nmol/h g FW for the control samples).

1.7.6.3   Fennel

Artes et al. (2002) in a series of experiments, looked into the treatment of fennel with ascorbic (1 %) and citric (5 %) acids, packed in PP baskets sealed with PP film to generate a MA and stored for 14 days at 0 °C followed by 4 days in air at 15 °C. The use of antioxidant solutions did not affect the weight loss with samples in OPP having a significantly lower weight loss compared to control (0.1 % and 3.5 to 3.9 % respectively) after cold storage. Flavor evaluation at harvest (8.2) revealed no changes after cold storage, but deteriorated significantly in both control (7.1) and OPP (7.3) after the 4-day period.

The effects of PMAP [OPP bags (treat. 1) and plastic baskets with OPP film (treat. 2)] to inhibit browning of the butt end cut zone of fennel bulbs stored over 14 days at 0 °C followed by complementary air storage during 3 days at 15 °C were studied. The color of bulbs stored in OPP bags was not negatively affected during complementary shelf life storage remaining in the cold storage levels. An increase by 3 to 4 % was in all treatments after complementary shelf life with no significant differences between treatments (Escalona et al. 2004).

Diced fennel washed in chlorinated water (100 mg/L) was stored under PMAP [OPP bags (treat. 1) and plastic baskets with OPP film (treat. 2)] at 0 °C for 14 days. Both treatments limited effectively the TPC growth (5.3 log CFU/g for treat. 1 and to 5.4 log CFU/g for treat. 2) compared to control samples (6.4 log CFU/g). A remarkable decrease in $L*$ value was recorded (from 72.8 at harvest to 60.8 for control and to 55.6 or 62.3 for samples stored in bags and baskets, respectively) for both treatments compared to initial values (Escalona et al. 2005).

1.7.6.4   Asparagus

The changes in lignifying, antioxidant enzyme activities and cell wall compositions of fresh-cut green asparagus in 1 mg/L aqueous ozone pretreated, and subsequent MAP (LDPE film) during storage at 3 °C for 25 days were investigated. For samples treated with O_3 and subsequent MAP storage, PAL activity had its highest value on the tenth day (0.33 U/mg) but at the end of storage was lower than the control (0.22 and 0.38 U/mg, respectively) (An et al. 2007).

An et al. (2006) treated asparagus [dipping in 20 ppm 6-benzylaminopurine (6-BA) for 10 min] and stored it under active (LDPE 25 μm film with 10 % O_2/5 % CO_2 initial atmosphere) and passive MAP (LDPE 15 μm film used) at 2 °C for 25 days. The use of 6-BA had a beneficial effect on chlorophyll content in both MAP applications (65 and 69 μg/g for passive and A MAP). The use of AMAP helped asparagus to retain the greatest lightness ($L*$), greenness (a*), hue angle (h°) and the least ΔE*.

Simon and Gonzalez-Fantos (2011) studied the changes in sensory and micro-biological quality of white asparagus stored under PMAP (2 OPP films, film A and film B with OTR: 13,200 and 45,000 mL/m² day atm. respectively) at 5 and 10 °C for up to 14 days. Shelf life was limited to 6 days at 10 °C. On samples stored at

5 °C, the use of film A proved beneficial with mesophile and *Enterobacteriaceae* counts being around 7 log/CFU/g at the end of storage and prolonging the shelf life up to 14 days.

Simon et al. (2004) evaluated the microbiological, quality and sensorial characteristics of white asparagus washed with chlorine or water and packaged under MAP (perforated PVC and P-Plus 160 films used) at 4 °C for 15 days. P-Plus had the lowest weight loss (0.2 %) while samples in PVC suffered higher losses (5.5 %). Water and chlorine treatment caused a reduction in mesophile and psychrotroph initial counts by 1 to 1.5 log cycles. *Enterobacteriaceae* counts in P-Plus packages were 1.6 logs higher on water washed samples (6.5 and 4.9 log CFU/g for water and chlorine treatments, respectively) (Table 1.4).

White asparagus spears were over-wrapped with a 16 mm stretch film (in 5 L glass jars with ethylene—free air passing through) and kept at 2.5, 5, 10, 15, 20 and 25 °C under darkness or light for 6 days. Ethanol in spear tissues was shown to increase linearly with storage temperature (100 and 800 μL/L at 2.5 and 20 °C, respectively). Weight loss of samples stored at 2.5–15 °C was less than 1 % while as temperature increased (20 °C and 25 °C) the losses amounted to 1.9 and 2.9 %, respectively. Spear quality was preserved in packages at 2.5 °C and 5 °C after 6 days of storage (Siomos et al. 2000).

Green asparagus spears were stored under refrigeration at 2 °C (first treatment), MAP (OPP P-Plus film was used) at 2 °C (second), and MAP at 10 °C after 5 days at 2 °C (third) until they were not suitable for consumption. The overall shelf-life was limited to 9–12, 26 and 14 days for the first, second and third treatment, respectively. Sear force of the basal segment increased during storage in all samples (70.2, 14.8 and 4.28 % for refrigerated samples, MAP samples at 10 °C and MAP samples at 2 °C, respectively) (Villanueva et al. 2005).

Asparagus spears treated with compressed argon and xenon (Ar/Xe 2/9 v:v) stored at 4 °C were compared to MAP (A MAP with 5 % O_2/5 % CO_2 initial atmosphere). MA packed samples had the lowest weight loss (0.5 %) after 18 days of storage. The use of Ar and Xe blocked the increase in cell permeability compared to control samples. Due to this effect, respiration peak and crude fiber formation was effectively restrained compared to MAP samples (Zhang et al. 2008).

1.7.6.5   Celery

LDPE and OPP were used to provide MA conditions for storage of celery sticks at 4 °C for 15 days. No significant changes in TSS (3.93°Brix for both films), TA (0.089 and 0.091 g citric acid/100 mL for LDPE and OPP) and pH (5.96 and 5.93 for LDPE and OPP) were recorded. The use of both films led to the products with an acceptable green color at the end of storage whereas samples stored under air displayed a 10 % decay rate (Gomez and Artes 2005).

Several CO_2 contents (0 %, 5 %, 10 %, 20 %, 30 %, 50 %) plus 2 % O_2 were used to preserve celeriac flakes at 4 °C and 15 °C (only for 0 %, 5 %, and 10 % CO_2 packs made from oriented polyamide/polyethylene laminated film). Mesophiles (not found

Table 1.4 A synoptical presentation of MAP application to asparagus

Species and food type	Initial gas mix	Package material	Treatment prior to packaging	Storage temperature (°C) and storage period (days)	Treatment effect and shelf life extension	References
Asparagus (*Asparagus officinalis* L.)	PMAP	Polypropylene P-Plus 35PA160		1. 2 °C for 26–33 days 2. 10 °C for 20 days	Samples stored under MAP at 2 °C were preserved for the time period of 28 days (191.7, 165.7, 354.3, 4 and 66.8 mg/kg for chlorophyll b, chlorophyll b′, chlorophyll a, pheophytin b and pheophytin a on the 28th day of storage)	Tenorio et al. (2004)
Asparagus (*Asparagus officinalis* L.) cv. UC800		LDPE film of 15 μm thickness and with OTR: 3.2 and CO_2TR: 10.2×10^{-12} mol s^{-1} mm^{-2} kPa^{-1}	Dipping in aqueous ozone solution (1 mg/L) for 30 min	3 °C for 25 days	Both enzyme activities and cell wall compositions were significantly reduced under ozone pretreatment and MAP storage (2 U/kg, 35 and 14 μmol/g for SOD, ascorbate peroxidase and glutathione reductase and 35, 43 and 58 mg/kg for lignin, cellulose and hemicellulose content on the 25th day). Enzyme activities remained high on MAP/ozone treated samples compared to control	An et al. (2007)

(continued)

Table 1.4 (continued)

Species and food type	Initial gas mix	Package material	Treatment prior to packaging	Storage temperature (°C) and storage period (days)	Treatment effect and shelf life extension	References
Asparagus (*Asparagus officinalis* L.)	1. PMAP 2. AMAP 10 kPa O_2/5 kPa O_2/ CO_2	1. LDPE film of 15 μm thickness and with OTR: 3.2 and CO_2TR: 10.2×10^{-12} mol s^{-1} mm^{-2} kPa^{-1} used for PMAP 2. LDPE film of 25 μm thickness and with OTR: 2.1 and CO_2TR: 6.5×10^{-12} mol s^{-1} mm^{-2} kPa^{-1} used for AMAP	Dipping in 20 ppm 6-benzylaminopurine (6-BA) for 10 min	2 °C for 24 days	MAP treated samples dipped in 6-BA retained their sensory attributes better compared to untreated samples, had bigger chlorophyll and ascorbic acid content and less fiber (69 μg/g, 24 mg/100 g and 0.38 % for total chlorophyll, ascorbic acid and crude fiber on the 24th day)	An et al. (2006)
Asparagus spears (*Asparagus officinalis* L.) cv Dariana	1. PMAP	Stretch film with OTR: 583 and CO_2TR: 1,750 mL/m^2 h atm.	Immersion in hot water (55 °C) for 3 min cooling in chlorinated water and then storage unpeeled or peeled	3 °C for 6 days	Heat treatment suppressed ethylene production of peeled asparagus (<1 ppm during the whole storage period). Pretreatment and MAP prevented lightness reduction and anthocyanins formation (81.43 and 0.9 μg/g for lightness and anthocyanin values of peeled treated samples)	Siomos et al. (2010)

Asparagus (*Asparagus officinalis* L.) cv. UC800	1. AMAP 5 %/5 % O_2/CO_2	PVC of 25 µm thickness with OTR: 3.6 and CO_2TR: 7.8×10^{-12} mol s^{-1} mm^{-2} kPa^{-1}	1. Compressed air treatment at 1.1 MPa for 24 h	4 °C for 18 days	Ar and Xe treated samples showed a shelf life of 12 days. MAP treated samples had minimal weight loss, both Ar-Xe treated and MAP samples had reduced crude fiber content (3–4 %) while chlorophyll was preserved better on both treatments (25 and 24 mg/100 g for MAP and Ar-Xe treated samples, respectively on the 18th day)	Zhang et al. (2008)
			2. Treatment with Ar-Xe mixture in 2:9 (v:v) partial pressure at 1.1 MPa for 24 h			
Asparagus (*Asparagus officinalis* L.) cv. Grolin	PMAP	1. OPP film with OTR: 13,200 mL/m^2 day atm. as film A	Washing with sodium hypochlorite solution containing 150 mg/L free chlorine. Samples were peeled	5 and 10 °C for 14 days	At 10 °C shelf life of the product was limited to 6 days while storage at a lower temperature with the use of A film resulted in a shelf life of 14 days (7 and 4.9 log CFU/g for TVC and *Enterobacteriaceae* on the 14th day)	Simon and Gonzalez-Fantos (2011)
		2. OPP film with OTR: 45,000 mL/m^2 day atm. as film B				

(continued)

Table 1.4 (continued)

Species and food type	Initial gas mix	Package material	Treatment prior to packaging	Storage temperature (°C) and storage period (days)	Treatment effect and shelf life extension	References
White asparagus (*Asparagus officinalis* L.) cv Cipres	PMAP	1. P-plus 160 film (OPP) of 35 μm thickness	Immersion in a sodium hypochlorite solution of 100 ppm concentration. Samples were peeled	4 °C for 15 days	The combination of antimicrobial washing and PMAP led to samples having minimal weight losses, total sugars of 2.8 %, ascorbic acid content of 110 mg/kg and lowered microbial counts (7, 7.4, 5, 1.8, 3.9 and 3.6 log CFU/g for TVC, psychrophiles, *Enterobacteriaceae*, coliform, LAB and aerobic spores on the 14th day of storage) being sensorially acceptable for 15 days	Simon et al. (2004)
Asparagus spears (*Asparagus officinalis* L.)	PMAP	16 μm stretch film of OTR: 583 and CO₂TR: 1,750 mL/ m² h atm.		2.5, 5, 10, 15, 20 and 25 °C under continuous light or darkness —6 days	Weight loss was significantly reduced by MAP (<1 % for temperatures under 15 °C). Ascorbic acid content remained close to the initial level (12 mg/100 g) while soluble solids of samples stored under 10 °C had values close to harvest levels (6.1 %)	Siomos et al. (2000)

White asparagus	AMAP 40 %/60 % CO_2/N_2	PP film of OTR: 44 and CO_2TR: 142 mL/m^2 day atm.	Samples were initially peeled and cut, washed in a chlorine solution of 30 ppm, immersed in an acidic bath of citric acid 0.2 % and ascorbic acid 0.1 %, drained, packaged under MAP and pasteurized	4, 10, 20 and 30 °C—17 days	Shelf life was correlated to storage temperature and was 84 h at 30 °C, 5 days at 20 °C, 6 days at 10 °C and 13 days at 4 °C. No *Clostridium* spp. was detected on samples stored at 4 and 10 °C. LAB populations on the time of rejection were (7.15, 5.85, 5.31 and 5.58 for samples stored at 4, 10, 20 and 30 °C, respectively)	Valero et al. (2006)
Asparagus (*Asparagus officinalis* L.) UC-157-F1	PMAP	OPP film with OTR: 14,000 mL/m^2 day atm. and of 35 µm thickness		1. 2 °C for 25 days 2. 2 °C for 5 days and then at 10 °C for the remaining days	MAP and refrigeration at 2 °C had the best results in preserving asparagus (2.1 %, 70.7 mg/100 g for weight loss and vitamin C content on the 33rd day), increasing its shelf life by 12 days compared to control and by 6 days compared to samples stored at 10 °C	Villanueva et al. (2005)

(continued)

Table 1.4 (continued)

Species and food type	Initial gas mix	Package material	Treatment prior to packaging	Storage temperature (°C) and storage period (days)	Treatment effect and shelf life extension	References
Asparagus (*Asparagus officinalis* L.)	PMAP	1. BOPP film used for control with OTR: 1,500 and CO_2TR: 3,280 mL/m² day atm. 2. BOPP film with breathable windows made from beta-PP films with OTR: 2,659,000 and CO_2TR: 2,014,000 mL/m² day atm. of different sizes (corresponding to 3 %, 6 %, 9 %, 12 %, 15 % and 30 % of the total surface area of the lidding film)	Dipping in 100 ppm Tsunami® 100 solution (peroxyacetic acid–based additive—an EPA-registered antimicrobial water additive) for 2 min	4 °C for 15 days	Shelf life of samples stored in packages made from the porous PP film had a shelf life of 29 days (for the films with the windows of 15 and 30 %) whereas control samples stored in packages with the BOPP film had a shelf life of <3 days	Chinsirikul et al. (2013)
Asparagus (*Asparagus officinalis* L.) var. UC-309	AMAP 8.15 kPa/17.51 kPa CO_2/O_2	OPP bags of 35 μm thickness and OTR: 3,500 mL/m² day atm.	Washing with chlorinated water containing 100 mg/L Cl_2 or with ozonated water at 0.1 mg/L O_3	4 °C for 28 days	Washing the samples with antimicrobial agents resulted in reduction of the microbial load (almost 2 and 4.5 log CFU/g for APC and *E. coli*, respectively). Shelf life of pretreated samples stored under MAP was determined to be 23 days	Sothornvit and Kiatchanapaibul (2009)

and 9.0×10^2 CFU/g for 5 and 10 % CO_2), psychrophiles (5.8×10^6 and 1.2×10^3 CFU/g for 5 and 10 % CO_2) and coliforms (3.0×10^1 and 5.8×10^2 CFU/g for 5 and 10 % CO_2) dropped under MAP at 4 °C (Radziejewska-Kubzdela et al. 2007).

1.7.7   Flowers

1.7.7.1   Artichoke

Five films (PVC, LDPE and three microperforated PP films PP_1, PP_2 and PP_3) were used to create MA conditions for artichokes stored at 5 °C for 8 days. A greater degree in water loss was found in control (1.7 %) and PVC (1.3 %) samples. Samples stored under low CO_2 atmospheres displayed increased phenolic compounds content compared to higher CO_2 packages (PP_1, PP_2 and PP_3) (Gil-Izquierdo et al. 2002).

1.7.7.2   Broccoli

MAP (PVC, LDPE 11, 15 and 20 μm were used) was applied to broccoli florets at 1 °C for 7 days and at 20 °C for 2.5 days to simulate the retail sale period. During shelf life, the weight losses were from 7 to 15 times lower in florets wrapped in LDPE compared to those wrapped in PVC. Artes et al. (2001) found, a minor trend to increase TA, generally not significant but remarkable in $LDPE_{20}$ (0.102 from 0.056 mg/100 mL at harvest), whereas the best results were obtained with $LDPE_{15}$ film.

Broccoli florets stored at 5 °C were monitored under the effect of PMAP (Xironet a 105 μm film), vent packaging (film with uniform perforations), and automatic misting (AM) by Barth and Zhuang (1996). Total carotenoids were preserved under MAP, but a loss (42–57 %) was observed in all other treatments by 6 days, however vitamin C was preserved best in MAP samples (−100 %) whereas control and VP samples retained only 14–46 %. A decrease in levels of moisture was recorded in VP samples (76 %) and both VP/AM and AM samples (84 %).

De Ell et al. (2006) studied the combination of sorbitol (water absorbent) and $KMnO_4$ (ethylene absorbent) in passive MAP (PD-961 film) for broccoli heads stored at 0–1 °C. The least amount of off-odor was found in broccoli in MAP containing 20 g of sorbitol. The weight loss of broccoli also enhanced with increasing content sorbitol (0.6–1.3 %). The amount of ethanol was greater in control bags (0.13 μL/L) and those with only $KMnO_4$ (0.12 μL/L), compared to the bags with sorbitol added (>0.6 μL/L).

Jacobsson et al. (2004a) analyzed the aroma compounds accumulated in broccoli stored at different passive MA [OPP, PVC, and LDPE with ethylene absorbing sachet at 10 °C for 7 days (1) or 4 °C for 3 days and 10 °C for the other 4 days (2)]. The original fresh weight diminished due to weight losses that varied between 0.6

and 1.6 %. Dimethylsulphide was detected in broccoli in case 2 (4.8–8.1 ppb), but not in case 1. In contrast, heat treatment increased aroma compounds as well as volatiles containing sulphur.

Broccoli stored under passive MA with OPP, LDPE with ethylene absorbing sachet and PVC at 10 °C for 7 days (1) or 4 °C for 3 days and 10 °C for the next 4 days (2). The materials under analysis revealed weight losses varied between 0.6 and 1.6 % of original fresh weight after 7 days of storage. The broccoli stored under condition 2, gave a more fresh impression than that of condition 1. Jacobsson et al. (2004b) showed that broccoli packaged in LDPE in the presence of an ethylene absorber displayed the most similar behavior to fresh.

The effect of passive MAP created by 4 films (OPP, PVC, 2 LDPE films one of which contained an ethylene absorber sachet) was studied by Jacobsson et al. (2004c) at 4 °C and 10 °C for 28 days. Although no chlorophyll degradation could be traced, in the broccoli a decrease in chlorophyll content was detected in broccoli stored at 10 °C. LDPE 2 maintained chlorophyll content better than the other packaging materials (25 % reduction at condition 2). Packaging prolonged the broccoli shelf-life up to 14 days (Table 1.5).

PE bags with no holes (M_0), 2 (M_1) and 4 (M_2) microholes were used to store broccoli florets at 4 and 20 °C. M_0 treatment maintained both total aliphatic and indole glucosinolates for 13 days, but decreased there after (26 and 15 % losses at 4 °C). The losses of aliphatic glucosinolates were 28 % and 15 % with M_2 and M_1 and the losses of indole glucosinolates amounted to 11 % and 8 %, respectively, for the same time period (13 days). Jia et al. (2009) demonstrated that the shelf life of broccoli florets stored under M_0 at either 4 or 20 °C was tripled.

The effects of packaging treatments (CA with 1.5 % O_2/6 % CO_2, passive MAP with microperforated LDPE on broccoli during storage time (10 and 25 days)) was investigated. Glucoraphanin in broccoli heads varied slightly during 25 days of storage under both air (9.2 µmol/g D/W) and CA (11.7 µmol/g D/W) at 4 °C. Rangkadilok et al. (2002) showed that in the CA, vegetables maintained their green color and freshness up to 25 days while yellowing (trace yellow) was observed only in broccoli heads stored under air.

Schreiner et al. (2007) monitored the effect of passive MAP [biaxial OPP with 2 (PMAP-1) and 8 (PMAP-2) microholes] on postharvest glucosinolate dynamics in mixed broccoli and cauliflower florets stored at 8 °C for 7 days. A minor decrease in the concentrations of glucoraphanin and glucoiberin at both treatments was observed (1.37, 0.28 and 1.43, 0.31 for PMAP-1 and PMAP-2, respectively with initial values 1.99 and 0.43).

Serrano et al. (2006) applied passive MAP with three types of film [macro (Ma-P), micro (Mi-P) and non-perforated (No-P) polypropylene] to broccoli at 1 °C for 28 days. Broccoli packaged with No-P and Mi-P films lost less than 1.5 % of their initial fresh weight, while this value increased up to 13.33 ± 0.60 % for broccoli in Ma-P film. Broccoli in Mi-P or No-P displayed a slight drop in texture over storage, with final force values being 116.68 ± 4.98 and 100.66 ± 7.58 N, respectively.

Broccoli buds were stored under passive MAP (PD-941 polymer film) or automatic misting (AM) treatment for 96 h at 5 °C and were studied by Zhuang et al. (1995).

Table 1.5 Effect of MAP and packaging material on broccoli shelf life prolongation

Species and food type	Initial gas mix	Packaging material	Treatment before packaging	Storage temperature (°C) and storage period (days)	Color	Microflora	Texture-weight loss	Sensory analysis	Shelf life(days)-life extension	References
Broccoli Florets (*Brassica oleracea*) var. Botrytis gr "Shogun" cv.	PMAP: 1. 18.3–15.5 % O₂/0.6–1.9 % CO₂ 2. 18–15.8 % O₂/0.8–2.5 % CO₂ 3. 18.7–15.3 % O₂/0.6–3.1 % CO₂ 4. 18.3–15.1 % O₂/0.9–3 % CO₂ For the first and second storage temperature	1. PVC 2. LDPE 11 μm 3. LDPE 15 μm 4. LDPE 20 μm		1 °C—7 days and 20 °C for 2.5 days	Samples stored in LDPE 15 and 20 showed the lowest browning disorders		Weight loss was significantly lower (7 to 15 times) in samples stored in LDPE films than for PVC film		LDPE 15 showed the best overall results and is considered the preferred film for broccoli florets storage	Artes et al. (2001)
Broccoli florets ("Iron Duke" cv.)	1. PMAP: 7.5 % O₂/11.2 % CO₂ 2. Vent packaging 3. Automatic Misting	1. PD-961 film 60 μm thick film for MAP 2. "Xtrone" 150 μm thick film for vent packaging		5 °C—6 days	Total carotenoids were preserved in MAP samples, whereas 42–57 % loss was observed in samples stored in other conditions		Moisture content was preserved in MAP treated samples, while there were decreased moisture levels (76 %) in vent packaged samples	MAP treated samples preserved initial color attributes better compared to both vent packaged samples and control	MAP was the best treatment, retaining vitamin C (14–46 % retention in other treatments), peroxidase activity and vitamin E content	Barth and Zhuang (1996)
Broccoli heads (*Brassica oleracea L*), *Italica* group, Marathon cv.	PMAP: 1.3–1.9 % O₂/8.9–10.4 % CO₂	PD-961 with OTR 6,000–8,000 and CDTR 19,000–22,000 mL/m² 24 h	1. KMnO₄ 20 g 2. 2.5 g Sorbitol + KMnO₄ 20 g 3. 5 g Sorbitol + KMnO₄ 20 g 4. 10 g Sorbitol + KMnO₄ 20 g 5. 20 g Sorbitol + KMnO₄ 20 g	0–1 °C—29 days			Weight losses were linearly increased with sorbitol increase, but their level did not exceed marketability limits	Addition of sorbitol reduced off-odors	The combination of sorbitol and KMnO₄ can play an important role in preserving quality of samples by reducing off-odors and off-flavors formation on broccoli	De Ell et al. (2006)

(continued)

Table 1.5 (continued)

Species and food type	Initial gas mix	Packaging material	Treatment before packaging	Storage temperature (°C) and storage period (days)	Color	Microflora	Texture-weight loss	Sensory analysis	Shelf life(days)-life extension	References
Broccoli (*Brassica oleracea* var. *Italica*) Marathon cv.	PMAP: 1. 14 % O_2/10.5 % CO_2 2. 6 % O_2/7 % CO_2 3. 17.9 % O_2/4 % CO_2	1. OPP with OTR 1.3 and CDTR 4.9 L/m^2 24 h atm. 2. LDPE with OTR 11 and CDTR >46 L/m^2 24 h atm. with ethylene adsorbing sachet 3. PVC with OTR 56.1 and CDTR >173 L/m^2 24 h atm.		1. 10 °C—7 days 2. 4 °C—3 days and 4 days at 10 °C			Weight loss incidence did not differ significantly between samples stored in the 3 packaging films (0.6–1.2 %)	Samples stored in OPP film had the most off odors, while samples in LDPE and PVC film preserved dimethylsulphide (0, 0 and 0 and 8.1, 2.4 and 4.8 for OPP, LDPE and PVC respectively, stored in temperatures 1 and 2), dimethyldisulphide (18, 5 and 4 and 26.6, 10.3 and 15.5 for OPP, LDPE and PVC respectively, stored in temperatures 1 and 2), and dimethyltrisulphide levels close to the initial values		Jacobsson et al. (2004a)
Broccoli (*Brassica oleracea* var. *Italica*) Marathon cv.	PMAP: 1. 14–12.9 % O_2/10.5–11.1 % CO_2 2. 6–4.2 % O_2/6.8–7.2 % CO_2 3. 17.9–16 % O_2/3.8–4.7 % CO_2 For the first and second storage temperature	1. OPP with OTR 1.3 and CDTR 4.9 L/m^2 24 h atm. 2. LDPE with OTR 11 and CDTR >46 L/m^2 24 h atm. with ethylene adsorbing sachet 3. PVC with OTR 56.1 and CDTR >173 L/m^2 24 h atm.		1. 10 °C—7 days 2. 4 °C—3 days and 4 days at 10 °C				Samples packaged in LDPE film had the highest values in freshness, greenness, compactness and evenness, compared	Overall, samples stored in LDPE bags with the addition of an ethylene absorber had properties close to the fresh samples	Jacobsson et al. (2004b)

Commodity	MAP/CA conditions	Packaging/film	Storage conditions	Result (quality)	Result (weight loss/texture)	Result (shelf-life/other)	Reference
Broccoli (*Brassica oleracea* var. *Italica*) "Monterey" cv.	PMAP: 1. 18.8–15.3 % O$_2$/2.5–6.7 % CO$_2$ 2. 13.2–12.6 % O$_2$/1.9–2.5 % CO$_2$ 3. 12.4–9.5 % O$_2$/2.9–4 % CO$_2$ 4. 20.6–18.8 % O$_2$/0.3–1.3 % CO$_2$ For the first and second storage temperature	1. OPP (OTR: 1,300 and CDTR: 4,900 mL/ m^2 day atm.) 2. LDPE with ethylene adsorbing sachet (OTR: 10,960 mL/ m^2 day atm.) 3. LDPE 2 (OTR: 4,290 and CDTR: 18,600 mL/ m^2 day atm.) 4. PVC (OTR: 56,100 and CDTR: >173,000 mL/ m^2 day atm.)	4 and 10 °C for 28 days	The longest shelf-life, when 30 % of the head had turned yellow, was achieved by OPP film (15 days at 4 °C and 10 days at 10 °C)	Weight losses were less in samples stored in OPP and LDPE 2 bags. Samples in LDPE 2 retained textural properties for 2 weeks at both storage temperatures	Samples were preserved for a longer time period stored at 4 °C. Shelf life of samples at 4 °C was 15, 12 and 11 days for samples stored in OPP, LDPE and LDPE 2 bags, respectively. At 10 °C, samples were acceptable for 7, 6 and 9 days stored in OPP, LDPE and LDPE 2 bags, respectively. LDPE 2 was the film that preserved overall quality at both temperatures	Jacobsson et al. (2004c)
Broccoli florets (*Brassica oleracea* var. *Italica*) "Youxiu" cv.	PMAP	1. PE with no holes (M$_0$) 2. PE with 2 holes (M$_1$) 3. PE with 4 holes (M$_2$)	Washing with a 50 ppm NaOCl solution for 1 min	4 and 20 °C—23 and 5 days, respectively	Visual quality was best retained by M$_0$ film, followed by M$_1$ and M$_2$ film	Weight losses were lower on MAP treated samples (<4 %) with the lowest values attained by M$_0$ film	M$_0$ treated samples retained acceptable visual attributes and glucosinolate (74 % of indole and 74 of total aliphatic glucosinolates at 4 °C and 78 % of total aliphatic glucosinolates contents at 20 °C) for 13 days at 4 °C and 3 days at 20 °C	Jia et al. (2009)
Broccoli heads (*Brassica oleracea* var. *Italica*) "Marathon" cv.	1. CA with 1.5 % O$_2$/6 % CO$_2$ 2. PMAP	1. LDPE without holes and stored at 4 °C 2. LDPE with microholes and stored at 20 °C Both films were used for passive MAP	4 and 20 °C—25 and 10 days, respectively	Visual quality of samples stored at 4 °C. MAP and control samples was preserved in both MAP and control samples. At 20 °C, acceptable visual quality was preserved for 7 and 3 days, for MAP and control samples, respectively		Glucoraphanin concentrations remained stable close to the initial levels on samples stored under PMAP at 4 °C (4.3 μmol/g)	Rangkadilok et al. (2002)

(continued)

Table 1.5 (continued)

Species and food type	Initial gas mix	Packaging material	Treatment before packaging	Storage temperature (°C) and storage period (days)	Color	Microflora	Texture-weight loss	Sensory analysis	Shelf life(days)-life extension	References
Broccoli florets cv. *Milady*	PMAP: 1. 1 % O_2/21 % CO_2 2. 8 % O_2/14 % CO_2	1. BOPP with 2 microholes 2. BOPP with 8 microholes		8 °C—7 days			Weight losses of both treatments were low and did not exceed 1.8 %	Off odors and visual quality was preserved by PMAP 1	PMAP 1 created by BOPP 1 film proved to be beneficial for preserving sensory attributes and glucosinolate content (1.65 and 3.56 for total aliphatic and idole glucosinolates, respectively) for 7 days at 8 °C	Schreiner et al. (2007)
Broccoli heads (*Brassica oleracea* L. var. *Italica*) Marathon cv.	PMAP: 1. 20 kPa O_2/0.08 kPa CO_2 2. 14 kPa O_2/2–2.5 kPa CO_2 3. 5 kPa O_2/6 kPa CO_2	1. Macroperforated (Ma-P) (OTR—1,600, CDTR—3,600) 2. Microperforated (Mi-P) (OTR—2,500, CDTR—25,000) 3. Nonperforated (No-P) (OTR—1,600, CDTR—3,600 mL/ m^2 day atm.)		1 °C—28 days			Weight loss of samples in Mi-P and No-P film s was limited >1.5 %, compared to 13.33 % of samples stored in Ma-P films. Texture values of Mi-P and No-P samples slightly decreased reaching 116, 68 and 100.66 N, respectively on the 28th day	No-P broccoli had no significant changes on color and chlorophyll content (3.5 from the initial 3.8 mg/g), while a slight decrease of both color attributes and chlorophyll was monitored on Mi-P samples (2.7 mg/g)	Samples stored in Mi-P and No-P film had extended shelf life reaching 28 days compared to 5 days of control samples	Serrano et al. (2006)

| Broccoli (Brassica oleracea L.) cv. Acadi | PMAP: 3 % O_2/8 % CO_2 | 26 L plastic containers with OTR: 21.76×10^{-12} mol/s pa and CDTR: 61.52 10^{-12} mol/s pa at 3 °C | 1. 3 °C—30 days 2. 3 °C—8 days, 13 °C—2 days, 3 °C—8 days, 13 °C—2 days. This pattern repeated for 30 storage days | Samples stored at 3 °C under PMAP had better color attributes and chlorophyll content (0.33 mg/g) | Samples stored at constant temperatures did not reveal signs of visible infection, whereas samples underwent temperature fluctuation had loss due to bacterial blotches (6.2 %) | Weight loss was significantly less on samples stored at constant temperature under MAP (2.9 %) | Color attributes and green color was preserved better on MAP treated samples stored at 3 °C without temperature fluctuation | Tano et al. (2007) |
| Broccoli heads (Brassica oleracea var. Italica) Iron Duke cv. | PMAP: 1. 8.7 % O_2/4 % CO_2 | Cryovac PD-941 | Non-packaged samples underwent automatic misting | 5 °C for 96 h | | | Samples stored under MAP retained chlorophyll content (2.35 µmol/g) and preserved polysaturated fatty acids (35.8) and lipoxygenase activity (3.2 µmol/min/g on DW basis) | Zhuang et al. (1995) |

The moisture of buds decreased down to 80 % for both treatments. However, no significant changes were recorded in Chl$a$, Chl$b$, and TChl of MAP-treated broccoli whereas for control samples about 50 % of Chl$a$ (1.0 pmol/g of DW), Chl$b$ (0.4 pmol/g of DW) and TChl (1.4 pmol/g of DW) were lost. In samples treated with AM, the protein level amounted to 85 mg/g of DW by 96 h that is lower by 20 % than the initial level.

1.7.7.3 Cauliflower

Simon et al. (2008) used two different films (non-perforated PVC and OPP) to generate atmosphere modification for cauliflowers stored at 4 °C and 8 °C for 20 days. Samples from PVC and PP bags had lower shear force values compared to control (2,520, 2,480 N for PVC and PP, respectively). Weight losses were greater for cauliflowers in PVC films than those stored in PP (2 and 0.61 % and 2.3 and 0.61 % at 4 °C and 8 °C, respectively). Samples in PP bags were given higher appearance scores.

Eight varieties of cauliflower (Abruzzi, Dulis, Casper, Serrano, Caprio, Nautilus, Beluga and Arbon) were stored under MAP (microperforated PVC, P-Plus 120, 160 and 240 films) at 4 °C for 25 days Sanz et al. (2007). The high initial microbial count was one of the reasons that samples from the Serrano variety in PVC film reached the microbial legal limit (7 log CFU/g) in only 3 days. P-Plus film was the most suitable film for maintaining sensory attributes in Abruzzi, Dulis, Casper and Serrano varieties.

1.7.8 Seeds

1.7.8.1 Mung bean

Chlorine dioxide (dipping in 100 ppm ClO_2 solution for 5 min) was applied in combination with MAP (passive MA with AVR 008 film used, vacuum and A MA with 100 % CO_2 and 100 % N_2) in mung bean sprouts during storage at 5 ± 2 °C for 7 days. Treatment with ClO_2 under vacuum, N_2 gas, and CO_2 gas led to reduction of total mesophiles (7.34, 7.51 and 7.22 $\log_{10}$ CFU/g respectively) at the end of storage. The use of ClO_2 as a disinfectant in conjunction with high CO_2 MAP was successful in inhibiting microbial growth (Jin and Lee 2007).

1.7.8.2 Snow Peas

The impact of precooling, PMAP [polymethyl pentene 25 μm (PMP-1) and 35 μm (PMP-2), LDPE and OPP were used] and CA (2.5, 5 and 10 kPa O_2 with 5 kPa CO_2, 0, 5 and 10 kPa CO_2 with 5 kPa O_2) storage on the preservation of snow pea pods at

5 °C was determined. PMP pods were the best and OPP the worst in terms of external appearance while precooling proved helpful. The application of CA with 5–10 kPa O_2 with 5 kPa CO_2 proved to be the most effective, since the changes in organic acid, free amino acid, sugar contents, and sensory attributes were slight (Pariasca et al. 2000).

1.7.8.3 Fungi

Truffles

Gamma irradiation (doses of 1.5 kGy and 2.5 kGy of either one) and subsequent PMAP (a 25 μm film was used) storage was used for *Tuber aestivum* truffles preservation. Treatment with 2.5 kGy irradiation led to doubling shelf life of samples (42 days) by restraining microbial populations (5.8, 5.1, >6, 2.8, and >2.4 log CFU/g reduction on mesophile, *Pseudomonas*, *Enterobacteriaceae*, LAB and yeasts counts, respectively compared to control) and preserving texture and aroma characteristics (Rivera et al. 2011a).

Treatments with electron beam and gamma irradiation (doses of 1.5 kGy and 2.5 kGy of either one) were tested on *Tuber melanosporum* truffles stored under MAP for 35 days at 4 °C. Mesophile growth was reduced by both treatments (4.3, 5.6, 6.4 and 6.6 log CFU/g for 1.5, 2.5 respectively). *Candida sake* and *Candida membranifaciens* var. *santamariae* survived and were the dominant microbial populations. The use of high irradiation dosages did not favor texture characteristics limiting shelf life to 28 days (Rivera et al. 2011b).

A shelf life of 28 and 21 days was achieved for *T. melanosporum* and *T. aestivum* truffles respectively stored at 4 °C under PMAP (a Mi-P film was used) microbial load was reduced (2 and 1.2 log CFU/g reduction for mesophiles, 3.7 and 1 log CFU/g reduction for *Enterobacteriaceae* and 2.8 and 0.8 log CFU/g reduction for Pseudomonas on *T. melanosporum* and *T. aestivum* samples, respectively stored under MAP when compared with air stored samples) and texture and sensory characteristics (aroma, color) where better preserved on MAP samples (Rivera et al. 2010).

A 4 log reduction for pseudomonads, more than 2 log reductions for *Enterobacteriacae*, moulds and LAB and 1.5 log reductions for yeasts on *Tuber aestivum* and *Tuber melanosporum* truffles surface was achieved by dipping in 70 % ethanol combined with 35 Hz ultrasound treatment. The subsequent storage under MAP (a microperforated LDPE film was used) at 4 °C led to a shelf life of 28 days for both truffles species (Rivera et al. 2011c).

Mushrooms

Antmann et al. (2008) investigated shiitake mushrooms stored under passive (two macroperforated PE films used, PE_A and PE_B) and active (15 % O_2 and 25 % O_2 in a LDPE film) atmosphere for 18 days at 5 °C. AMAP samples manage to retain their initial weight through the whole storage period while both control and PMAP samples had significant losses (>10 % and <5 % weight loss for films A and B respectively). Mushrooms in packages B had a shelf life of 10 days due to their sensory deterioration, particularly by changes in the color and uniformity of their gills.

A macroperforated (control) PP, a LDPE and a PP film were used for MA packaging of shiitake mushrooms at 5 °C for 16 days. Control samples had a weight loss that reached 47.5 % whereas samples stored in PP or PE exhibited a weight loss of 5.6 %. The use of MAP reduced the estimated shelf life reaching 5 ± 2 days for mushrooms stored in PE or PP, and 12 ± 2 days for ma-PP for a 25 % consumer rejection (Ares et al. 2006).

Combinations of antimicrobial (ClO_2 and H_2O_2 were tested in various concentrations and washing duration) and antibrowning (sodium d-isoascorbate and H_2O_2 in various concentrations) treatments were tested on sliced mushrooms stored under PMAP (PA-190 OPP was used) at 4 and 8 °C for 7 days. The use of ClO_2 had satisfactory antimicrobial effect. H_2O_2 washing for 60 s produced best quality (higher $L*$ values) and (>1 log CFU reduction) was superior to ClO_2 in reducing *pseudomonad* counts (Cliffe-Byrnes and O'Beirne 2008).

The effect of a biobased packaging material (paper coated with a wheat gluten solution) on the quality of common mushrooms stored at 20 °C was examined by Guillaume et al. (2010). Shelf life of samples was extended by 2 days, while the main disadvantage of the material was a significant weight loss observed (3.8 % on day 3).

High oxygen atmosphere (70, 80 and 95 % O_2 in a barrier film) and PMAP (bioriented PP film used) were compared for preserving sliced mushrooms at 4 °C for 7 days. Mushrooms stored under low O_2 atmospheres had a reduced shelf life (by 3 days) compared to samples under high O_2 MA (6 days). High O_2 levels had a beneficial effect on product acceptability (7 days, while EMA samples were rejected on the sixth day due to off-odors) (Jacxsens et al. 2001).

Oyster mushrooms were washed with water, 0.5 % citric acid, 0.5 % calcium chloride, and 0.5 % citric acid with 0.5 % calcium chloride and stored under PMAP (PP, 0.075, 0.05 and 0.0375 mm LDPE and LLDPE films were used) at 8 °C for 8 days. Mushrooms packaged in 0.015-mm LLDPE after washing with 0.5 % calcium chloride and 0.5 % citric acid produced the least off odors and had the least off-color development (Jayathunge and Illeperuma 2005).

Shiitake mushrooms were stored under PMAP (an LDPE film used with OTR: $1{,}078 \times 10^{-18}$ mol/m s Pa) after UV-C irradiation treatment and were stored at 1 ± 1 °C for 16 days and finally held at 20 °C for 3 days. UV-C treated samples retained higher ascorbic acid and flavonoid levels (32 and 22 mg/kg at day 16 respectively) while the decrease in firmness was reduced (Jiang et al 2010a).

An atmosphere modification of 12 O_2 and 2.5 CO_2 and PMAP with the use of LDPE film (OTR: $1{,}078 \times 10^{-18}$ mol/m s Pa) was used by Jiang et al. (2010b) for preserving *Agaricus bisporus* mushrooms for 15 days at 4 ± 1 °C. The use of a protective atmosphere reduced browning by inhibiting POD activity and the accumulation of lignin (3.5 2.3 and 2.2 $\times 103$A280/kg for control, AMAP and PMAP samples respectively).

Jiang et al. (2010c) investigated the effect of gamma irradiation (1.0, 1.5 and 2 kGy) on shiitake mushrooms stored under PMAP (BOPP film was used) at 4 °C for 20 days. Mushrooms irradiated with 1.0 kGy maintained firmness, phenolic and flavonoid compounds and had a shelf life of 20 days.

Jiang et al. (2010d) investigated the structural changes of the cell wall of shiitake mushrooms stored under PMAP (LDPE film used with 0, 2 and 4 microholes) at 4 °C for 16 days. The film with 4 microholes proved effective in maintaining firmness, reducing losses in protein (46 %) and retarded cellulose increase (35 g/kg DW).

Agaricus bisporus mushrooms were dipped in 2.2-(hydroxynitrosohydrazino)-bisethanamine (DETANO), a nitric oxide donor (0.5, 1, and 2 mM concentrations) and subsequently stored under MAP (BOPP film used) and stored at 4 °C for 16 days. Firmness was better maintained with 1 mM DETANO and MAP and the use of DETANO led to a lower browning incidence. The use of NO combined with MAP can lead to shelf life extension up to 12 days (Jiang et al. 2011).

Enoki mushrooms were packaged under various conditions (full and half vacuum with RD-106 polyolefin for packaging film, half vacuum with cast polypropylene and polyolefins RD-106 and PD-941 for films and RD-106 packages stored at 5, 10 and 15 °C) and stored at 10 °C for 14 days. The higher the degree of vacuum the lower the weight loss the samples had (1.3, 1.1 and 1 % for air, partial and fully vacuumized samples). A higher degree of initial vacuum in RD-106 film packages suppressed stripe elongation (6.5 %) (Kang et al. 2000).

Sliced and whole mushrooms were spray-coated with solution of chitosan and $CaCl_2$ (2 g per 100 mL) and packaged with a PVC wrap or two polyolefins (PD-941 and PD-961) at 12 °C for 6 days. Coated sliced mushrooms exhibited higher ΔE than uncoated mushrooms while the use of PD-961 film had the best color results in terms of L^* value. Products coated with chitosan were less mature at the end of the experiment compared to uncoated ones (Kim et al. 2006).

$CaCl_2O_2$ (0.4 and 0.8 g/L) treatment and passive (LDPE with 2 or 4 perforations or PVC) or active atmosphere modification (10 mg/100 mL O_2/10 mg/100 mL CO_2) was tested for the preservation of mushrooms stored at 5 ± 1 °C for 10 days. Coliforms [6×10^6, 7×10^4 and 4×10^5 CFU/g for treated LDPE (4 per.) with MAP, PVC and perforated PVC] and the total plate counts [10^6, 10^5 and 4×10^5 CFU/g for treated LDPE (4 per.) with MAP, PVC and perforated PVC] of all packaging methods were lower with the addition of 0.8 g/L $CaCl_2O_2$ (Kuyper et al. 1993).

Pseudomonas fluorescens and *Candida sake* were inoculated into homogenized mushrooms and stored under different gaseous atmospheres [CO_2/O_2 (25 %/1 %)-MAP_1 and CO_2/O_2 (50 %/1 %)-MAP_2] at 5 and 10 °C for 18 days. PH was greatly reduced in MAP_2 storage conditions (5.06 and 5.79 for MAP_2 at 5 and 10 °C respectively were the lowest pH values). The presence of CO_2 played an important role in increasing the lag time of *Pseudomonas* and *Candida* (Masson et al. 2002).

Figure 1.4 show that the use of LDPE as a packaging film and treatment with calcium hypochlorite had a beneficial effect on retention of *pseudomonads*. The combination of hydrogen peroxide and sodium isoascorbate reduced the initial microbial load but did not have the same effect in the population's growth rate.

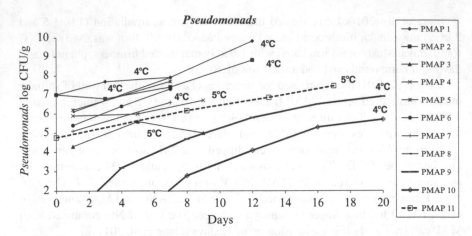

Fig. 1.4 *Pseudomonas* growth on mushrooms treated and stored under modified atmosphere conditions [PMAP 1 and 2 with microperforated oriented polypropylene used as packaging film (with 45,000 and 2,400 mL/m² day atm. O₂ permeabilities respectively), Simon et al. (2005), for PMAP 3 mushrooms treated with $CaCl_2O_2$ spray (0.4 mg/mL) were packed in perforated LDPE bags, for PMAP 4 perforated LDPE was the packaging film, Kuyper et al. (1993), washing with ClO_2 (50 mg/mL) for 60 s and stored under PMAP 5 (PA-160 OPP film used), washing with H_2O_2 3 % for 60 s and then stored under PMAP 6 (PA-160 used), washing into a H_2O_2 (3 %) and sodium isoascorbate (4 %) solution and stored under PMAP 7 (PA-160 used) and PMAP 8 was created with the use of PA-160 film after washing with H_2O_2 (3 %) for 60 s and spraying the products with sodium isoascorbate (4 %), Cliffe-Byrnes and O'Beirne (2008). Shiitake mushrooms were stored under PMAP 9 and PMAP 10 after gamma irradiation with a dose of 1.0 and 2.0 kGy, respectively (Jiang et al. 2010a, b, c, d) PMAP 11 was prepared with the use of a PVC film (12 μm thickness, OTR: 25,000 mL/m² day atm.) after immersion of the *Agaricus bisporus* mushrooms in an aqueous solution with 10 g/L of citric acid for 5 min (Simon et al. 2010)

Parentelli et al. (2007) used a combination of passive (LDPE and PP films) and active (5 % O_2/2.5 % CO_2) atmosphere conditions for storage of shiitake mushrooms at 5 °C for 20 days. MAP retained weight loss under 6.5 % at the end of storage. The use of atmosphere modification, both passive and active, increased deterioration rate and deteriorated sensory attributes, compared to control samples.

Roy et al. (1996) used sorbitol and sodium chloride (5, 10 or 15 g) in an attempt to modify the in-package relative humidity (IPRH) of mushrooms (water irrigated and $CaCl_2$ irrigated) stored in MAP (polyethylene film) at 12 °C. The addition of sorbitol kept IPRH over 80 % during the storage period. Packages containing $CaCl_2$ irrigated mushrooms had lower IPRH than normally grown mushrooms. NG mushrooms stored with 10 and 15 g sorbitol had higher $L*$ values and lower ΔE values than without sorbitol.

A research conducted led to the conclusions that *Agaricus bisporus* mushrooms can safely be stored for 32, 44 and 108 h under passive MAP conditions with the use of LDPE, PP and OPP films respectively.

Simon et al. (2005) applied MA conditions that were created passively (PVC, OPP_1 and OPP_2 films with high and low O_2 permeability respectively used) for the storage of sliced mushrooms at 4 ± 1 °C for 13 days. The increase in the shear force needed was bigger in the control and PVC samples (1,020 N), while there was less force needed for the shearing of samples stored in PP_1 (890 N) and PP_2 (780 N) bags. The use of PP_2 bags did not prevent the detection of off-odors, although the samples had the best appearance scores (no deformation, blotch absence).

A reduction by 2.5 log CFU/g was achieved with the immersion of *Agaricus bisporus* mushrooms in an aqueous citric acid solution (10 g/L) and subsequent storage under passive MAP (PVC film used with OTR: 25,000 mL/m^2 day atm.). Shelf life of samples washed with the citric acid solution and stored under PMAP was 13 days due to microbial spoilage (Simon et al 2010).

Mushrooms were stored in plastic containers (creating modified atmosphere conditions) and were subjected to temperature fluctuations (TF) (4 and 14 °C for 2 days alternatively) during the 12-day storage period. The RH levels were temperature depended and with each temperature increase RH dropped rapidly (90 %) and returning to 100 % when temperature was reaching the initial value. Due to TF, there was a high ethanol content increase (24.4-fold) in MA packages compared to ones with stable temperature conditions (1.9-fold) (Tano et al. 2007).

The effects of vacuum cooling prior to packaging under MAP (5 ± 1 % O_2/3 ± 1 % CO_2 initial atmosphere with LDPE film used) or hypobaric conditions (20–30 kPa total pressure) at 4 ± 1 °C were evaluated by Tao et al. (2006). Weight loss was restrained under MAP whereas it was relatively extensive on mushrooms under HC (>1 % and 14.78 %, respectively). The membrane permeability of mushrooms under MAP was 15 %.

VC and AMAP (5 ± 1 % O_2/3 ± 1 % CO_2 initial atmosphere with LDPE film used) at 4 ± 1 °C were applied and well defined by Tao et al. (2007). The activities of superoxide dismutase, catalase, peroxidase and polyphenol oxidase were positively affected by vacuum cooling resulting in a 1.2-, 1.2-, 1.1- and 1.1-fold increase, respectively. Browning was more apparent in control samples than in MA packaged and VC treated ones (1.5° of browning).

Villaescusa and Gil (2003) tested various storage temperatures (0, 4 and 7 °C) MA conditions [PVC, two microperforated polypropylenes (MPP_1 and MPP_2) and a LDPE film were used for storage at 4 °C] and different moisture absorbers (10, 15 and 20 g of sorbitol and 3, 5, 7, and 15 g of silica gel in MPP_2 film) for prolonging the shelf life of mushrooms. The optimal temperature for mushroom storage was 0 °C since all the quality characteristics remained close to the initial levels (2.0 % weight loss, 4.2°Brix SSC, 6.2 pH and 0.09 g citric acid/100 mL TA).

The use of superatmospheric conditions (50, 70, 80, 90 and 100 % O_2) for the preservation of needle mushrooms (PE film, 10 g of bentonite as a moisture absorber and 8 g of active carbon as a deodorant were used) at 3 °C for 34 days was examined by Wang et al. (2011). Mushrooms under 80 % O_2 had the best appearance scores, the lowest POD and PPO levels (4.26 and 3.25 U/g respectively) and the highest SOD levels (0.73 U/g %) leading to the lowest browning levels and an increased shelf life.

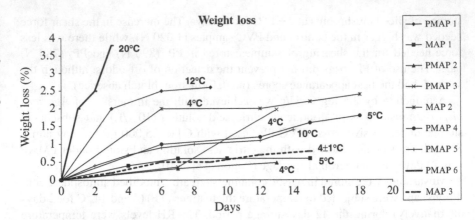

Fig. 1.5 Weight loss in mushrooms stored under passive or active modified atmosphere conditions [PMAP 1 was created in polyolefin pouches with OTR: 13,800 mL/m² day atm., Roy et al. (1996), MAP 1 with initial atmosphere of 5 ± 1 % O_2/3 ± 1 % CO_2 and LDPE film used after vacuum cooling treatment at 5 °C, Tao et al. (2006), PMAP 2 and 3 with microperforated OPP used as packaging film (with 45,000 and 2,400 mL/m² day atm. O_2 permeabilities, respectively), Simon et al. (2005), MAP 2 was created in PE packages with initial conditions of 15 % O_2, a macroperforated PE (17 perforations/m² with 0.1 mm² surface) was used for shiitake mushrooms storage at PMAP 4, Antmann et al. (2008), for PMAP 5 an PO film was used to preserve fresh enoki mushrooms, Kang et al. (2000), for common mushrooms stored under PMAP 6 a paper film coated with a wheat gluten solution was used (Guillaume et al. 2010) and MAP 3 for *Agaricus bisporus* mushrooms stored under MAP (12 O_2 and 2.5 CO_2) by Jiang et al. (2010a, b, c, d)]

An evaluation of Fig. 1.5 leads to the conclusion that packaging under passive or active atmosphere conditions has beneficial effects on weight loss (all samples but one were under 2.5 %). Storage under low temperatures and the selection of films with low water vapor transmission rate can reduce water loss and increase shelf life of the product.

1.7.9 Other

1.7.9.1 Coleslaw mix (80 % Shredded Cabbage, 20 % Shredded Carrots)

Coleslaw mix inoculated with *L. monocytogenes* and *L. innocua* and then stored under PMAP (OPP and four microperforated OPP films, PA-120, PA-160, PA-190 and PA-210) at 3 and 8 °C was studied by Bourke and O'Beirne (2004). By day 12, *L. innocua* counts fell rapidly in PA-120 (1.7 log CFU/g) than in all other PA films. The initial record of approximately 10^6 CFU/g total aerobic mesophiles at 3 °C increased by 2 more log cycles by day 7 and finally remained at this level in the PA films, while OPP counts declined.

Different packaging treatments (PMAP with OPP and four microperforated OPP films, PA-120, PA-160, PA-190 and PA-210) were used to determine the effects of different packaging treatments on the quality of dry coleslaw stored at 4 and 8 °C for 9 days (Cliffe-Byrnes et al. 2003). The rate of deterioration on sensory scores was significantly greater at 8 °C with PA films scoring significantly higher than OPP. Weight loss was considerably higher in air and statistically greater at 8 °C. Although it may not seem to be the best option, packaging within microperforated films allowed better preservation of quality.

The effect of chlorine treatment (washing in a 100 ppm chlorine solution for 5 min) and PMAP (microperforated OPP films PA-160 and PA-210 used) on coleslaw mix stored at 4 and 8 °C for 9 days was monitored by Cliffe-Byrnes and O'Beirne (2005). Chlorine washed coleslaw at 4 °C remained within sensory acceptable limits throughout the storage period. Chlorine also reduced the initial counts of *Pseudomonas* species, at both 4 (7.05 and 7.04 log CFU/g) and 8 °C (7.3 and 7.33 log CFU/g) for both films (PA-160 and 210). PA-160 with chlorine washing at 4 °C was found to be the best combination.

Coleslaw mix inoculated with *L. monocytogenes* and two strains of *E. coli* and stored afterwards under passive MAP (OPP film) at 4 and 8 °C for 12 days was analyzed by Francis and O'Beirne (2001). *L. monocytogenes* numbers was shown to decrease by 1.5 log cycles during storage at 8 °C, and by approximately 2.0 log cycles when samples were held at 4 °C. Populations of *E. coli O157:H7* on coleslaw mix increased by approximately 1.5 log cycles by day 5, whereas they declined by 1 to 2 log cycles, depending on strain, with further storage at 8 °C.

Francis et al. (2007) focused on the contribution of the glutamate decarboxylase (GAD) acid resistance system to survival and growth of *L. monocytogenes* LO28 in PMAP (OPP 35 μm) coleslaw stored at 4, 8 and 15 °C for 12 days. A wild strain of *L. monocytogenes* and 4 strains with mutant genes (ΔgadA, ΔgadB, ΔgadC, ΔgadAB) of negligible GAD activity were inoculated on the coleslaw. At 4 °C, the populations of the ΔgadAB (2.7 log CFU/g) were lower than the wild-type or other strains (about 4.1 log). At 15 °C, the order of survival was LO28 > ΔgadA > ΔgadC > ΔgadB > ΔgadAB.

1.8 Conclusions

From the previous work, the ability of different packaging conditions to delay the quality loss of the vegetables was explained. Factors such as storage temperature, use of several pretreatments, film permeability or light and dark storage conditions combined with atmosphere modification led to shelf life extension. Shelf life extension achieved on chicory endive varied from 2 to 3 days. A 14 to 60-day storage life was observed in tomato samples stored under MAP. Carrots stored under MAP remained edible from a time period of 7 to 15 days, while broccoli under MAP gained a 7 to 14-day storage period elongation. For 7 to 14 days mushrooms that underwent MAP storage were sensorially acceptable, while 28 days was the shelf life limit for asparagus in MA packages. Postharvest quality of peppers was maintained for 15 to 28 days by altering the in-package atmosphere and lettuce preservation reached 10 days under MAP. A summary referring on shelf life of vegetables stored under MAP is shown on Table 1.6.

Table 1.6 Effect of MAP storage on shelf life of fresh vegetables

Commodity	Type of plastic	Atmospheric composition		Storage temperature	Storage period	References
		% O_2	% CO_2			
Artichoke	LDPE with OTR: 2.1 and CDTR: 4.6×10^{-10} mol/s m² Pa	7.7 (EMA)	9.8 (EMA)	5 °C	8 days	Gil-Izquierdo et al. (2002)
Asparagus	OPP with OTR: 13,200 and CDTR: 10,000 mL/m² day atm.	14.5–17.5 (EMA)	5.6–7.8 (EMA)	5 °C	14 days	Simon and Gonzalez-Fantos (2011)
Bamboo shoots	LDPE	2	5	10 °C	10 days	Shen et al. (2006)
Betel leaves	PP film with OTR: 1.49 and CDTR: 5.24×10^{-6} mL m/m²/h/kPa	9	6.5–7	20 °C	10 days	Rai et al. (2010)
Broccoli	OPP with OTR: 1,300 and CDTR: 4,900 mL/m² day atm.	15.3–18.8 (EMA)	2.5–6.7 (EMA)	4 °C	15 days	Jacobsson et al. (2004c)
Broccoli florets	PE	2 (EMA)	14 (EMA)	4 °C	13 days	Jia et al. (2009)
Broccoli heads	PP with OTR: 1,600 and CDTR: 3,600 mL/m² day atm.	5 (EMA)	6 (EMA)	1 °C	28 days	Serrano et al. (2006)
Cabbage, shredded		5.8 (EMA)	4.2 (EMA)	4 °C	9 days	Gomez-Lopez et al. (2007b)
Carrots	PE	5	5	5 °C	10 days	Alasalvar et al. (2005)
Carrots	Film with OTR: 3,529 mL O_2/kg h	4.5 % (EMA)	8.9 (EMA)	7 °C	8 days	Gomez-Lopez et al. (2007a)
Cauliflower	OPP with OTR: 45,000 and CDTR: 45,000 mL/m² day atm.	15–18 (EMA)	4 (EMA)	4 °C	20 days	Simon et al. (2008)
Celery sticks	OPP with OTR: 5,500 and CDTR: 10,000 mL/m² day atm.	6 (EMA)	7 (EMA)	4 °C	15 days	Gomez and Artes (2005)
Chicory endive	OTR: 3,704 mL O_2/m² 24 h atm.	3	2–5	7 °C	6 days	Jacxsens et al. (2003)
Cucumber	LDPE	16–17 (EMA)	3–4 (EMA)	5 °C	12 days	Wang and Qi (1997)

	HDPE film	0	30	10 °C	14 days	Arvanitoyannis et al. (2005)
Eggplant, grafted						
Fennel, diced	OPP with OTR: 5,500 and CDTR: 10,000 mL/m² day atm.	11–13 (EMA)	9–12 (EMA)	0 °C	14 days	Escalona et al. (2005)
Garlic, sprouts	PVC with OTR: 5,500 and CDTR: 10,000 mL/m² day atm.	5.8 (EMA)	6.5 (EMA)	4 °C	15 days	Li et al. (2010)
Ginseng	PVC film with OTR: 235 10⁻⁵ mL/mm² h atm.	8–15 (EMA)	6–13 (EMA)	0 °C	210 days	Hu et al. (2005)
Kale	PVC film with OTR: 12,889 mL/m² day	9–15 (EMA)	4–10 (EMA)	1 °C	17 days	Kobori et al. (2011)
Kohlrabi sticks	Amide-PE	7 (EMA)	9 (EMA)	0 °C	14 days	Escalona et al. (2007a)
Lettuce, Iceberg	OPP film with OTR: 2,000 mL/m² day atm.	1.5 (EMA)	12 (EMA)	4 °C	8 days	Pirovani et al. (1997)
Lettuce, Lollo Rosso	BOPP with OTR: 1,800 mL/m² day atm.	6 (EMA)	4 (EMA)	5 °C	8 days	Allende and Artes (2003b)
Lettuce, Romaine	LDPE with OTR: 5,676 and CDTR: 29,435 mL O₂/m² 24 h bar	9 (EMA)	3 (EMA)	0 °C	14 days	Manolopoulou et al. (2010)
Lotus	PE and chitosan coating	8.9 (EMA)	6.9 (EMA)	4 °C	10 days	Xing et al. (2010)
Mushrooms, Agaricus sliced.	PO (PD-961) with OTR: 7,000 and CDTR: 21,500 mL/m² day atm.	2 (EMA)	3 (EMA)	12 °C	6 days	Kim et al. (2006)
Mushrooms, Enoki	PO (RD-106) with OTR: 166.3 and CDTR: 731.2 mL/m² h atm.	1.7–3.4 (EMA)	2.6–5.8 (EMA)	10 °C	8 days	Kang et al. (2000)
Onions, sliced	LDPE with OTR: 15 mL/m² day atm.	1	40	-2 °C	12 days	Liu and Li (2006)
Pepper	PD-961 with OTR: 6,000–8,000 mL/m² 24 h atm.	1–4 (EMA)	6 (EMA)	5 °C	21 days	Gonzalez-Aguilar et al. (2004)
Potato	PD-PO film with OTR: 16,000 and CDTR: 36,000 mL/m² 24 h atm.	2.1	6	2 °C	21 days.	Gunes and Lee (1997)

(continued)

Table 1.6 (continued)

Potato, cubes	PP film with OTR: 2,000–2,500 mL/ m² 24 h atm.			4±1 °C	4 weeks	Baskaran et al. (2007)
Pumpkin, cut	LDPE	2 (EMA)	18 (EMA)	5±2 °C	25 days	Habibunnisa et al. (2001)
Rocket	OPP film with OTR: 1,500 mL/ m² day	5	10	5±1 °C	10 days	Arvanitoyannis et al. (2011a, b)
Snow peas	PMP with OTR: 79.4×10^{-12} mol/s m² Pa	5 (EMA)	5 (EMA)	5 °C	28 days	Pariasca et al. (2000)
Sweet potato slices	PD-961 PO film with OTR: 7,000 and CDTR: 21,000 mL/m² 24 h atm.	3 (EMA)	4 (EMA)	2 °C	14 days	Erturk and Picha (2007)
Sweet potato, shredded	PD-961 PO film with OTR: 7,000 and CDTR: 21,000 mL/m² 24 h atm.	5	4	4 °C	14 days	McConnell et al. (2005)
Tomato, slices-wedges	OPP with OTR: 5,500 and CDTR: 10,000 mL/m² 24 h atm.	3	0–4	0 °C	14 days	Aguayo et al. (2004)
Truffles, *T. aestivum*	Microperforated LDPE-Polyester film	7 (EMA)	15 (EMA)	4 °C	21 days	Rivera et al. (2010)
Truffles, *T. melanosporum*	Microperforated LDPE-Polyester film	7 (EMA)	15 (EMA)	4 °C	28 days	Rivera et al. (2010)
Zucchini, sliced	OPP with OTR: 653 and CDTR: 1,222 mL/m² 24 h atm.	5	5	5 °C	6–7 days	Lucera et al. (2010)

References

Abadias M, Alegre I, Oliveira M, Altisent R, Vinas I (2012) Growth potential of *Escherichia coli* *O157-H7* on fresh-cut fruits (melon and pineapple) and vegetables (carrot and escarole) stored under different conditions. Food Control 27:37–44

Abdul-Raouf UM, Beuchat LR, Ammar MS (1993) Survival and growth of *Escherichia coli* *O15:H7* on salad vegetables. Appl Environ Microbiol 59:1999–2006

Aghdama MS, Sevillano L, Flores FB, Bodbodak S (2013) Heat shock proteins as biochemical markers for postharvest chilling stress in fruits and vegetables. Sci Hortic 160:54–64

Aguayo E, Escalona V, Artes F (2004) Quality of fresh-cut tomato as affected by type of cut, packaging, temperature and storage time. Eur Food Res Technol 219:492–499

Ahn H-J, Kim J-H, Kim J-K, Kim D-H, Yook H-S, Byun M-W (2005) Combined effects of irradiation and modified atmosphere packaging on minimally processed Chinese cabbage (*Brassica rapa L.*). Food Chem 89:589–597

Akbudak B, Akbudak N, Seniz V, Eris A (2012) Effect of pre-harvest harpin and modified atmosphere packaging on quality of cherry tomato cultivars "Alona" and "Cluster". Br Food J 114(2):180–196

Alasalvar C, Al-Farsi M, Quantick PC, Shahidi F, Wiktorowicz R (2005) Effect of chill storage and modified atmosphere packaging (MAP) on antioxidant activity, anthocyanins, carotenoids, phenolics and sensory quality of ready-to-eat shredded orange and purple carrots. Food Chem 89:69–76

Allende A, Artes F (2003a) Combined ultraviolet-C and modified atmosphere packaging treatments for reducing microbial growth of fresh processed lettuce. LWT 36:779–786

Allende A, Artes F (2003b) UV-C radiation as a novel technique for keeping quality of fresh processed "*Lollo Rosso*" lettuce. Food Res Int 36:739–746

Allende A, Aguayo E, Artes F (2004a) Microbial and sensory quality of commercial fresh processed red lettuce throughout the production chain and shelf life. Int J Food Microbiol 91:109–117

Allende A, Luo Y, McEvoy JL, Artés F, Wang CY (2004b) Microbial and quality changes in minimally processed baby spinach leaves stored under super atmospheric oxygen and modified atmosphere conditions. Postharvest Biol Technol 33:51–59

Amanatidou A, Slump RA, Gorris LGM, Smid EJ (2000) High oxygen and high carbon dioxide modified atmospheres for shelf-life extension of minimally processed carrots. J Food Sci 65(1):61–66

An J, Zhang M, Lu Q, Zhang Z (2006) Effect of a prestorage treatment with 6-benzylaminopurine and modified atmosphere packaging storage on the respiration and quality of green asparagus spears. J Food Eng 77:951–957

An J, Zhang M, Lu Q (2007) Changes in some quality indexes in fresh-cut green asparagus pretreated with aqueous ozone and subsequent modified atmosphere packaging. J Food Eng 78:340–344

Antmann G, Ares G, Lema P, Lareo C (2008) Influence of modified atmosphere packaging on sensory quality of *shiitake* mushrooms. Postharvest Biol Technol 49:164–170

Ares G, Parentelli C, Gambaro A, Lareo C, Lema P (2006) Sensory shelf life of *shiitake* mushrooms stored under passive modified atmosphere. Postharvest Biol Technol 41:191–197

Ares G, Lareo C, Lema P (2008) Sensory shelf life of butterhead lettuce leaves in active and passive modified atmosphere packages. Int J Food Sci Technol 43:1671–1677

Artes F, Conesa MA, Hernandez S, Gil MI (1999) Keeping quality of fresh-cut tomato. Postharvest Biol Technol 17:153–162

Artes F, Vallejo F, Martinez JA (2001) Quality of broccoli as influenced by film wrapping during shipment. Eur Food Res Technol 213:480–483

Artes F, Escalona VH, Artes-Hdez F (2002) Modified atmosphere packaging of fennel. J Food Sci 67(4):1550–1554

Arvanitoyannis SI, Bouletis DA (2012) Minimally processed vegetables in "Modified atmosphere and active packaging technologies" CRC press. Taylor and Francis Group, Boca Raton

Arvanitoyannis IS, Khah EM, Christakou EC, Bletsos FA (2005) Effect of grafting and modified atmosphere packaging on eggplant quality parameters during storage. Int J Food Sci Technol 40:311–322

Arvanitoyannis IS, Bouletis DA, Papa AE, Gkagtzis CD, Hadjichristodoulou C, Papaloukas C (2011a) Microbial and sensory quality of "Lollo verde" lettuce and rocket salad stored under active atmosphere packaging. Anaerobe 17:307–309

Arvanitoyannis IS, Bouletis DA, Papa AE, Gkagtzis CD, Hadjichristodoulou C, Papaloukas C (2011b) The effect of addition of olive oil and "Aceto balsamico di Modena" wine vinegar in conjunction with active atmosphere packaging on the microbial and sensory quality of "Lollo Verde" lettuce and rocket salad. Anaerobe 17:303–306

Ayhan Z, Esturk O, Tas E (2008) Effect of modified atmosphere packaging on the quality and shelf life of minimally processed carrots. Turk J Agric For 32:57–64

Bailen G, Gullen F, Castillo S, Serrano M, Valero D, Martinez-Romero D (2006) Use of activated carbon inside modified atmosphere packages to maintain tomato fruit quality during cold storage. J Agric Food Chem 54:2229–2235

Barry-Ryan C, O'Beirne D (1999) Ascorbic acid retention in shredded iceberg lettuce as affected by minimal processing. J Food Sci 64(3):498–500

Barry-Ryan C, Pacussi JM, O'Beirne D (2000) Quality of shredded carrots as affected by packaging film and storage temperature. J Food Sci 65(4):726–730

Barth MM, Zhuang H (1996) Packaging design affects antioxidant vitamin retention and quality of broccoli florets during postharvest storage. Postharvest Biol Technol 9:141–150

Baskaran R, Usha DA, Nayak CA, Kudachikar VB, Prakash MNK, Prakash M, Ramana KVR, Rastogi NK (2007) Effect of low-dose γ-irradiation on the shelf life and quality characteristics of minimally processed potato cubes under modified atmosphere packaging. Radiat Phys Chem 76:1042–1049

Batu A, Thompson AK (1998) Effects of modified atmosphere packaging on post-harvest qualities of pink tomatoes. Turk J Agric For 22:365–372

Beltran D, Selma MV, Marian A, Gil MI (2005a) Ozonated water extends the shelf life of fresh-cut lettuce. J Agric Food Chem 53:5654–5663

Beltran D, Selma MV, Tudela JA, Gil MI (2005b) Effect of different sanitizers on microbial and sensory quality of fresh-cut potato strips stored under modified atmosphere or vacuum packaging. Postharvest Biol Technol 37:37–46

Bennik MHJ, Peppelenbos HW, Nguyen-the C, Carlin F, Smid EJ, Gorris LGM (1996) Microbiology of minimally processed, modified-atmosphere packaged chicory endive. Postharvest Biol Technol 9:209–221

Beuchat LR, Brakett RE (1990) Inhibitory effects of raw carrots on Listeria monocytogenes. Appl Environ Microbiol 56(6):1734–1742

Bidawid S, Farber JM, Sattar SA (2001) Survival of hepatitis A virus on modified atmosphere-packaged (MAP) lettuce. Food Microbiol 18:95–102

Bourke P, O'Beirne D (2004) Effects of packaging type, gas atmosphere and storage temperature on survival and growth of Listeria spp. in shredded dry coleslaw and its components. Int J Food Sci Technol 39:509–523

Brecht JK, Chau KV, Fonseca SC, Oliveira FAR, Silva FM, Nunes MCN, Bender RJ (2003) Maintaining optimal atmosphere conditions for fruits and vegetables throughout the postharvest handling chain. Postharvest Biol Technol 27:87–101

Caleb JO, Opara LU, Witthuhn RC (2012) Modified atmosphere packaging of pomegranate fruit and arils: a review. Food Bioprocess Technol 5:15–30

Caleb JO, Mahajan VP, Al-Said AF, Opara LU (2013) Modified atmosphere packaging technology of fresh and fresh-cut produce and the microbial consequences—a review. Food Bioprocess Technol 6:303–329

Carlin F, Nguyen-the C, Da Silva AA, Cochet C (1996) Effects of carbon dioxide on the fate of Listeria monocytogenes, of aerobic bacteria and on the development of spoilage in minimally processed fresh endive. Int J Food Microbiol 32:159–172

Charles F, Guillaume C, Gontard N (2008) Effect of passive and active modified atmosphere packaging on quality changes of fresh endives. Postharvest Biol Technol 48:22–29

Chavarri MJ, Herrera A, Arino A (2005) The decrease in pesticides in fruit and vegetables during commercial processing. Int J Food Sci Technol 40:205–211

Chinsirikul W, Klintham P, Kerddonfag N, Winotapun C, Hararak B, Kumsang P, Chonhenchob V (2013) Porous ultrahigh gas-permeable polypropylene film and application in controlling in-pack atmosphere for asparagus. Packag Technol Sci. doi:10.1002/pts.2027

Chua D, Goh K, Saftner RA, Bhagwat AA (2008) Fresh-cut lettuce in modified atmosphere packages stored at improper temperatures supports enterohemorrhagic *E. coli* isolates to survive gastric acid challenge. J Food Sci 73(9):M148–M153

Cliffe-Byrnes V, O'Beirne D (2005) Effects of chlorine treatment and packaging on the quality and shelf-life of modified atmosphere (MA) packaged coleslaw mix. Food Control 16:707–716

Cliffe-Byrnes V, O'Beirne D (2007) The effects of modified atmospheres, edible coating and storage temperatures on the sensory quality of carrot discs. Int J Food Sci Technol 42:1338–1349

Cliffe-Byrnes V, O'Beirne D (2008) Effects of washing treatment on microbial and sensory quality of modified atmosphere (MA) packaged fresh sliced mushroom (*Agaricus bisporus*). Postharvest Biol Technol 48:283–294

Cliffe-Byrnes V, Mc Laughlin CP, O'Beirne D (2003) The effects of packaging film and storage temperature on the quality of a dry coleslaw mix packaged in a modified atmosphere. Int J Food Sci Technol 38:187–199

Das E, Gurakan GC, Bayındırlı A (2006) Effect of controlled atmosphere storage, modified atmosphere packaging and gaseous ozone treatment on the survival of *Salmonella Enteritidis* on cherry tomatoes. Food Microbiol 23:430–438

De Ell JR, Toivonen PMA, Cornut F, Roger C, Vigneault C (2006) Addition of sorbitol with $KMnO_4$ improves broccoli quality retention in modified atmosphere packages. J Food Qual 29:65–75

Dhall RK (2013) Advances in edible coatings for fresh fruits and vegetables: a review. Crit Rev Food Sci Nutr 53(5):435–450

Erturk E, Picha DH (2007) Effect of temperature and packaging film on nutritional quality of fresh-cut sweet potatoes. J Food Qual 30:450–465

Escalona VH, Aguayo E, Gomez P, Artes F (2004) Modified atmosphere packaging inhibits browning in fennel. LWT 37:115–121

Escalona VH, Aguayo E, Artes F (2005) Overall quality throughout shelf life of minimally fresh processed fennel. J Food Sci 70(1):S13–S17

Escalona VH, Verlinden BE, Geysen S, Nicolaı BM (2006) Changes in respiration of fresh-cut butterhead lettuce under controlled atmospheres using low and superatmospheric oxygen conditions with different carbon dioxide levels. Postharvest Biol Technol 39:48–55

Escalona VH, Aguayo E, Artes F (2007a) Quality changes of fresh-cut kohlrabi sticks under modified atmosphere packaging. J Food Sci 72(5):S303–S307

Escalona VH, Aguayo E, Artes F (2007b) Extending the shelf life of kohlrabi stems by modified atmosphere packaging. J Food Sci 72(5):S308–S313

Escalona VH, Aguayo E, Artes F (2007c) Modified atmosphere packaging improved quality of kohlrabi stems. LWT 40:397–403

Fan X, Sokorai JBK (2011) Effects of gamma irradiation, modified atmosphere packaging, and delay of irradiation on quality of fresh-cut iceberg lettuce. Hortscience 46(2):273–277

Fan X, Toivonen PMA, Rajkowski KT, Sokorai KJB (2003) Warm water treatment in combination with modified atmosphere packaging reduces undesirable effects of irradiation on the quality of fresh-cut iceberg lettuce. J Agric Food Chem 51:1231–1236

FDA/CFSAN (2001) Analysis and evaluation of preventive control measures for the control and reduction/elimination of microbial hazards on fresh and fresh-cut produce. http://www.fda.gov/Food/ScienceResearch/ResearchAreas/SafePractices for Food Processes/ucm091368.htm

Francis GA, O'Beirne D (1997) Effects of gas atmosphere, antimicrobial dip and temperature on the fate of *Listeria innocua* and *Listeria monocytogenes* on minimally processed lettuce. Int J Food Sci Technol 32:141–151

Francis GA, O'Beirne D (2001) Effects of vegetable type, package atmosphere and storage temperature on growth and survival of *Escherichia coli O157:H7* and *Listeria monocytogenes*. J Ind Microbiol Biotechnol 27:111–116

Francis GA, Thomas C, O'Beirne D (1999) The microbiological safety of minimally processed vegetables. Int J Food Sci Technol 34:1–22

Francis GA, Scollard J, Meally A, Bolton DJ, Gahan CGM, Cotter PD, Hill C, O'Beirne D (2007) The glutamate decarboxylase acid resistance mechanism affects survival of *Listeria monocytogenes LO28* in modified atmosphere-packaged foods. J Appl Microbiol 103:2316–2324

Gil MI, Ferreres F, Tomas-Barberan FA (1999) Effect of postharvest storage and processing on the antioxidant constituents (flavonoids and vitamin C) of fresh-cut spinach. J Agric Food Chem 47:2213–2217

Gil MI, Conesa MA, Artes F (2002) Quality changes in fresh cut tomato as affected by modified atmosphere packaging. Postharvest Biol Technol 25:199–207

Gil-Izquierdo A, Conesa MA, Ferreres F, Gil MA (2002) Influence of modified atmosphere packaging on quality, vitamin C and phenolic content of artichokes (*Cynara scolymus L.*). Eur Food Res Technol 215:21–27

Gleeson E, O'Beirne D (2005) Effects of process severity on survival and growth of *Escherichia coli* and *Listeria innocua* on minimally processed vegetables. Food Control 16:677–685

Gomez PA, Artes F (2005) Improved keeping quality of minimally fresh processed celery sticks by modified atmosphere packaging. LWT 38:323–329

Gomez-Lopez VM, Devlieghere F, Bonduelle V, Debevere J (2005) Intense light pulses decontamination of minimally processed vegetables and their shelf-life. Int J Food Microbiol 103:79–89

Gomez-Lopez VM, Devlieghere F, Ragaert P, Debevere J (2007a) Shelf-life extension of minimally processed carrots by gaseous chlorine dioxide. Int J Food Microbiol 116:221–227

Gomez-Lopez VM, Ragaert P, Ryckeboer J, Jeyachchandran V, Debevere J, Devlieghere F (2007b) Shelf-life of minimally processed cabbage treated with neutral electrolysed oxidising water and stored under equilibrium modified atmosphere. Int J Food Microbiol 117:91–98

Gomez-Lopez VM, Ragaert P, Jeyachchandran V, Debevere J, Devlieghere F (2008) Shelf-life of minimally processed lettuce and cabbage treated with gaseous chlorine dioxide and cysteine. Int J Food Microbiol 121:74–83

Gonzalez-Aguilar GA, Ayala-Zavala JF, Ruiz-Cruz S, Acedo-Felix E, Diaz-Cinco ME (2004) Effect of temperature and modified atmosphere packaging on overall quality of fresh-cut bell peppers. LWT 37:817–826

Guillaume C, Schwab I, Gastaldi E, Gontard N (2010) Biobased packaging for improving preservation of fresh common mushrooms (*Agaricus bisporus L.*). Innovative Food Sci Emerg Technol 11:690–696

Gunes G, Lee CY (1997) Color of minimally processed potatoes as affected by modified atmosphere packaging and antibrowning agents. J Food Sci 62(3):572–575

Habibunnisa BR, Prasad R, Shivaiah KM (2001) Storage behaviour of minimally processed pumpkin (*Cucurbita maxima*) under modified atmosphere packaging conditions. Eur Food Res Technol 212:165–169

Hagenmaier RD, Baker RA (1997) Low-dose irradiation of cut iceberg lettuce in modified atmosphere packaging. J Agric Food Chem 45:2864–2868

Harris LJ, Farber JN, Beuchat LR, Parish ME, Suslow TV, Garrett EH, Busta FF (2003) Outbreaks associated with fresh produce-Incidence, growth, and survival of pathogens in fresh and fresh-cut produce. Comprehensive Rev Food Sci Food Safety 2(Suppl S1):78–141

Heaton JC, Jones K (2008) Microbial contamination of fruit and vegetables and the behavior of enteropathogens in the phyllosphere: a review. J Appl Microbiol 104:613–626

Hong JH, Gross KC (2001) Maintaining quality of fresh-cut tomato slices through modified atmosphere packaging and low temperature storage. J Food Sci 66(7):960–965

Hong S-I, Kim D (2004) The effect of packaging treatment on the storage quality of minimally processed bunched onions. Int J Food Sci Technol 39:1033–1041

Hu W, Xu P, Uchino T (2005) Extending storage life of fresh ginseng by modified atmosphere packaging. J Sci Food Agric 85:2475–2481

Irtwange VS (2006) Application of modified atmosphere packaging and related technology in postharvest handling of fresh fruits and vegetables. *Agric Eng Int CIGR eJ Invited Overv* 4(8)

Islam E, Xiao-e Y, Zhen-li H, Qaisar M (2007) Assessing potential dietary toxicity of heavy metals in selected vegetables and food crops. J Zhejiang Univ Sci B 8(1):1–13

Jacobsson A, Nielsen T, Sjoholm I (2004a) Influence of temperature, modified atmosphere packaging, and heat treatment on aroma compounds in broccoli. J Agric Food Chem 52:1607–1614

Jacobsson A, Nielsen T, Sjoholm I (2004b) Effects of type of packaging material on shelf-life of fresh broccoli by means of changes in weight color and texture. Eur Food Res Technol 218: 157–163

Jacobsson A, Nielsen T, Sjoholm I, Werdin K (2004c) Influence of packaging material and storage condition on the sensory quality of broccoli. Food Qual Prefer 15:301–310

Jacxsens L, Delvieghere F, Van Der Steen C, Debevere J (2001) Effect of high oxygen modified atmosphere packaging on microbial growth and sensorial qualities of fresh-cut produce. Int J Food Microbiol 71:197–210

Jacxsens L, Devlieghere F, Debevere J (2002a) Temperature dependence of shelf-life as affected by microbial proliferation and sensory quality of equilibrium modified atmosphere packaged fresh produce. Postharvest Biol Technol 26:59–73

Jacxsens L, Devlieghere F, Debevere J (2002b) Predictive modelling for packaging design equilibrium modified atmosphere packages of fresh-cut vegetables subjected to a simulated distribution chain. Int J Food Microbiol 73:331–341

Jacxsens L, Devlieghere F, Ragaert P, Vanneste E, Debevere J (2003) Relation between microbiological quality, metabolite production and sensory quality of equilibrium modified atmosphere packaged fresh cut produce. Int J Food Microbiol 83:263–280

Jayathunge L, Illeperuma C (2005) Extension of postharvest life of oyster mushroom by modified atmosphere packaging technique. J Food Sci 70(9):E573–E578

Jeon BS, Lee CY (1999) Shelf-life extension of American fresh ginseng by controlled atmosphere storage and modified atmosphere packaging. J Food Sci 64(2):328–331

Jia C-G, Wei C-J, Wei J, Yan G-F, Wang B-L, Wang Q-M (2009) Effect of modified atmosphere packaging on visual quality and glucosinolates of broccoli florets. Food Chem 114(1):28–37

Jiang T, Jahangir MM, Jiang Z, Lu X, Ying T (2010a) Influence of UV-C treatment on antioxidant capacity, antioxidant enzyme activity and texture of postharvest shiitake (*Lentinus edodes*) mushrooms during storage. Postharvest Biol Technol 56:209–215

Jiang T, Jahangir MM, Wang Q, Ying T (2010b) Accumulation of lignin and malondialdehyde in relation to quality changes of button mushrooms (*Agaricus Bisporus*) stored in modified atmosphere. Food Sci Technol Int 16(3):217–224

Jiang T, Luo S, Chen Q, Shen L, Yin T (2010c) Effect of integrated application of gamma irradiation and modified atmosphere packaging on physicochemical and microbiological properties of shiitake mushroom (*Lentinus edodes*). Food Chem 122:761–767

Jiang T, Wang Q, Xu S, Jahangir MM, Ying T (2010d) Structure and composition changes in the cell wall in relation to texture of shiitake mushrooms (*Lentinula edodes*) stored in modified atmosphere packaging. J Sci Food Agric 90:742–749

Jiang T, Zheng X, Li J, Jing G, Cai L, Ying T (2011) Integrated application of nitric oxide and modified atmosphere packaging to improve quality retention of button mushroom (*Agaricus bisporus*). Food Chem 126:1693–1699

Jin H-H, Lee S-Y (2007) Combined effect of aqueous chlorine dioxide and modified atmosphere packaging on inhibiting *Salmonella typhimurium* and *Listeria monocytogenes* in mungbean sprouts. J Food Sci 72(9):M441–M445

Kakiomenou K, Tassou C, Nychas G-J (1998) Survival of *Salmonella enteritidis* and *Listeria monocytogenes* on salad vegetables. World J Microbiol Biotechnol 14:383–387

Kang J-S, Park W-P, Lee D-S (2000) Quality of *enoki* mushrooms as affected by packaging conditions. J Sci Food Agric 81:109–114

Karakas B, Yildiz F (2007) Peroxidation of membrane lipids in minimally processed cucumbers packaged under modified atmospheres. Food Chem 100:1011–1018

Kaur P, Rai RD, Paul S (2011) Quality changes in fresh-cut spinach (*Spinacia oleracea*) under modified atmospheres with perforations. J Food Qual 34:10–18

Kim JG, Luo Y, Tao Y, Saftner RA, Gross KC (2005) Effect of initial oxygen concentration and film oxygen transmission rate on the quality of fresh-cut romaine lettuce. J Sci Food Agric 85:1622–1630

Kim KM, Ko JA, Lee JS, Park HJ, Hanna MA (2006) Effect of modified atmosphere packaging on the shelf-life of coated, whole and sliced mushrooms. LWT 39:364–371

Kobori CN, Huber IS, Sarantopoulos CIGL, Rodriguez-Amaya DB (2011) Behavior of flavonols and carotenoids of minimally processed kale leaves during storage in passive modified atmosphere packaging. J Food Sci 76(2):H31–H37

Koide S, Shi J (2007) Microbial and quality evaluation of green peppers stored in biodegradable film packaging. Food Control 18:1121–1125

Koukounaras A, Siomos SA, Sfakiotakis E (2010) Effects of degree of cutting and storage on atmosphere composition, metabolic activity and quality of rocket leaves under modified atmosphere packaging. J Food Qual 33:303–316

Kuyper L, Weinert LAG, McGill AEJ (1993) The effect of modified atmosphere packaging and addition of calcium hypochlorite on the atmosphere composition, color and microbial quality of mushrooms. LWT 26(1):14–20

Lacroix M, Lafortune R (2004) Combined effects of gamma irradiation and modified atmosphere packaging on bacterial resistance in grated carrots (*Daucus carota*). Radiat Phys Chem 71:77–80

Lee S-Y, Baek S-Y (2008) Effect of chemical sanitizer combined with modified atmosphere packaging on inhibiting *Escherichia coli O157:H7* in commercial spinach. Food Microbiol 25:582–587

Lee KS, Woo KL, Lee DS (1994) Modified atmosphere packaging for green chili peppers. Packag Technol Sci 7:51–58

Lee H-H, Hong S-I, Kim D (2011) Microbiological and visual quality of fresh-cut cabbage as affected by packaging treatments. Food Sci Biotechnol 20(1):229–235

Li X, Li L, Wang X, Zhang L (2010) Improved keeping quality of fresh-cut garlic sprouts by atmosphere packaging conditions. In: Environmental science and information application technology (ESIAT), 2nd conference on environmental science and information application technology, p 317–320, ISBN 978-1-4244-7387-8

Lin D, Zhao Y (2007) Innovations in the development and application of edible coatings for fresh and minimally processed fruits and vegetables. Comprehensive Rev Food Sci Food Safety 6(3):60–75

Liu F, Li Y (2006) Storage characteristics and relationships between microbial growth parameters and shelf life of MAP sliced onions. Postharvest Biol Technol 40:262–268

Lucera A, Costa C, Mastromatteo M, Conte A, Del Nobile MA (2010) Influence of different packaging systems on fresh-cut zucchini (*Cucurbita pepo*). Innovative Food Sci Emerg Technol 11:361–368

Macura D, McCannel AM, Li MZC (2001) Survival of *Clostridium botulinum* in modified atmosphere packaged fresh whole North American ginseng roots. Food Res Int 34:123–125

Mae N, Makino Y, Oshita S, Kawagoe Y, Tanaka A, Akihiro T, Akama K, Koike S, Matsukura C, Ezura H (2010) Stimulation of γ-aminobutyric acid production in vine-ripened tomato (*Solanum lycopersicum* L.) fruits under an adjusted aerobic atmosphere. J Packag Sci Technol 19(5):375–381

Majidi H, Minaei S, Almasi M, Mostofi Y (2011) Total soluble solids, titratable acidity and repining index of tomato in various storage conditions. Aust J Basic Appl Sci 5(12):1723–1726

Mangaraj S, Goswami TK, Mahajan PV (2009) Applications of plastic films for modified atmosphere packaging of fruits and vegetables: a review. Food Eng Rev 1(2):133–158

Manolopoulou H, Gr L, Chatzis E, Xanthopoulos G, Aravantinos E (2010) Effect of temperature and modified atmosphere packaging on storage quality of fresh-cut romaine lettuce. J Food Qual 33:317–336

Martinez JA, Artes F (1999) Effect of packaging treatments and vacuum-cooling on quality of winter harvested iceberg lettuce. Food Res Int 32:621–627

Martinez I, Ares G, Lema P (2008) Influence of cut and packaging film on sensory quality of fresh-cut butterhead lettuce (*Lactuca sativa L., cv. Wang*). J Food Qual 31:48–66

Martinez-Sanchez A, Tudela AJ, Luna C, Allende A, Gil IM (2011) Low oxygen levels and light exposure affect quality of fresh-cut Romaine lettuce. Postharvest Biol Technol 59:34–42

Masson Y, Ainsworth P, Fuller D, Bozkurt H, Ibanoglu S (2002) Growth of *Pseudomonas fluorescens* and *Candida sake* in homogenized mushrooms under modified atmosphere. J Food Eng 54:125–131

Mastromatteo M, Conte A, Del Nobile MA (2012) Packaging strategies to prolong the shelf life of fresh carrots (*Daucus carota L.*). Innovative Food Sci Emerg Technol 13:215–220

McConnell RY, Truong V-D, Walter WM Jr, McFeeters RF (2005) Physical, chemical and microbial changes in shredded sweet potatoes. J Food Process Preserv 29:246–267

Merino S, Rubires X, Knochel S, Tomas JM (1995) Emerging pathogens, *Aeromonas spp.* Int J Food Microbiol 28:157–168

Nguyen-the C, Carlin F (1994) The microbiology of minimally processed fresh fruits and vegetables. Crit Rev Food Sci Nutr 34(4):371–401

Niemira BA, Boyd G (2013) Influence of modified atmosphere and varying time in storage on the irradiation sensitivity of *Salmonella* on sliced roma tomatoes. Radiat Phys Chem 90:120–124

Niemira BA, Fan X, Sokorai KJB (2005) Irradiation and modified atmosphere packaging of endive influences survival and regrowth of *Listeria monocytogenes* and product sensory qualities. Radiat Phys Chem 72:41–48

Odriozola-Serrano I, Soliva-Fortuny R, Martin-Belloso O (2008) Effect of minimal processing on bioactive compounds and color attributes of fresh-cut tomatoes. LWT 41:217–226

Oliveira M, Usall J, Solsona C, Alegre I, Vinas I, Abadias M (2010) Effects of packaging type and storage temperature on the growth of foodborne pathogens on shredded 'Romaine' lettuce. Food Microbiol 27:375–380

Ooraikul B, Stiles ME (1990) Modified atmosphere packaging of food. E. Horwood (Ed.). Chapman and Hall, London, pp 167–176

Parentelli C, Ares G, Corona M, Lareo C, Gambaro A, Soubes M, Lema P (2007) Sensory and microbiological quality of *shiitake* mushrooms in modified-atmosphere packages. J Sci Food Agric 87:1645–1652

Pariasca JAT, Miyazaki T, Hisaka H, Nakagawa H, Sato T (2000) Effect of modified atmosphere packaging (MAP) and controlled atmosphere (CA) storage on the quality of snow pea pods (*Pisum sativum L. var. saccharatum*). Postharvest Biol Technol 21:213–223

Phillips CA (1996) Review: modified atmosphere packaging and its effects on the microbiological quality and safety of produce. Int J Food Sci Technol 31:463–479

Piagentini AM, Guemes DR, Pirovani ME (2003) Mesophilic aerobic population of fresh-cut spinach as affected by chemical treatment and type of packaging film. J Food Sci 68(2):602–607

Pilon L, Oetterer M, Gallo CR, Spoto MHF (2006) Shelf life of minimally processed carrot and green pepper. Ciênc Tecnol Aliment Campinas 26(1):150–158

Pirovani ME, Guemes DR, Piagentini AM, Di Pentima JH (1997) Storage quality of minimally processed cabbage packaged in plastic films. J Food Qual 20:381–389

Pirovani ME, Piagentini AM, Guemes DR, Di Pentima JH (1998) Quality of minimally processed lettuce as influenced by packaging and chemical treatment. J Food Qual 22:475–484

Puligundla P, Jung J, Ko S (2012) Carbon dioxide sensors for intelligent food packaging applications. Food Control 25:328–333

Radziejewska-Kubzdela E, Czapski J, Czaczyk K (2007) The effect of packaging conditions on the quality of minimally processed celeriac flakes. Food Control 18:1191–1197

Rai RD, Chourasiya KV, Jha NS, Wanjari DO (2010) Effect of modified atmospheres on pigment and antioxidant retention of betel leaf (*Piper Betel L.*). J Food Biochem 34:905–915

Rangkadilok N, Tomkins B, Nicolas ME, Premier RR, Bennett RN, Eagling DR, Taylor PWJ (2002) The effect of post-harvest and packaging treatments on glucoraphanin concentration in broccoli (*Brassica oleracea var. italica*). J Agric Food Chem 50:7386–7391

Rico D, Martin-Diana AB, Barat MJ, Barry-Ryan C (2007) Extending and measuring the quality of fresh-cut fruit and vegetables: a review. Trends Food Sci Technol 18:373–386

Rivera SC, Blanco D, Salvador LM, Venturini EM (2010) Shelf-life extension of fresh *Tuber aestivum* and *Tuber melanosporum* truffles by modified atmosphere packaging with microperforated films. J Food Sci 75(4):E226–E233

Rivera SC, Blanco D, Marco P, Oria R, Venturini EM (2011a) Effects of electron-beam irradiation on the shelf life, microbial populations and sensory characteristics of summer truffles (*Tuber aestivum*) packaged under modified atmospheres. Food Microbiol 28:141–148

Rivera SC, Venturini EM, Marco P, Oria R, Blanco D (2011b) Effects of electron-beam and gamma irradiation treatments on the microbial populations, respiratory activity and sensory characteristics of *Tuber melanosporum* truffles packaged under modified atmospheres. Food Microbiol 28(7):1252–1260

Rivera SC, Venturini EM, Oria R, Blanco D (2011c) Selection of a decontamination treatment for fresh *Tuber aestivum* and *Tuber melanosporum* truffles packaged in modified atmospheres. Food Control 22:626–632

Rodriguez-Hidalgo S, Artes-Hernandez F, Gomez AP, Fernandez JA, Artes F (2010) Quality of fresh-cut baby spinach grown under a floating trays system as affected by nitrogen fertilisation and innovative packaging treatments. J Sci Food Agric 90:1089–1097

Roy S, Anantheswaran RC, Beelman RB (1996) Modified atmosphere and modified humidity packaging of fresh mushrooms. J Food Sci 61(2):391–397

Sabir KF, Agar IT (2011) Effects of 1-methylcyclopropene and modified atmosphere packing on postharvest life and quality in tomatoes. J Food Qual 34:111–118

Saltveit ME (2000) Wound induced changes in phenolic metabolism and tissue browning are altered by heat shock. Postharvest Biol Technol 21:61–69

Sandhya (2010) Modified atmosphere packaging of fresh produce: current status and future needs. LWT Food Sci Technol 43:381–392

Sanz S, Olarte C, Echavarri JF, Ayala F (2007) Evaluation of different varieties of cauliflower for minimal processing. J Sci Food Agric 87:266–273

Sayed Ali M, Nakano K, Maezawa S (2004) Combined effect of heat treatment and modified atmosphere packaging on the color development of cherry tomato. Postharvest Biol Technol 34:113–116

Schafer KS, Kegley SE (2002) Persistent toxic chemicals in the US food supply. J Epidemiol Community Health 56:813–817

Schreiner M, Peters P, Krumbein A (2007) Changes of glucosinolates in mixed fresh-cut broccoli and cauliflower florets in modified atmosphere packaging. J Food Sci 72(8):S585–S589

Serrano M, Martinez-Romero D, Guillen F, Castillo S, Valero D (2006) Maintenance of broccoli quality and functional properties during cold storage as affected by modified atmosphere packaging. Postharvest Biol Technol 39:61–68

Sharma M, Lakshman S, Ferguson S, Ingram TD, Luo Y, Patel J (2011) Effect of modified atmosphere packaging on the persistence and expression of virulence factors of *Escherichia coli O157-H7* on shredded iceberg lettuce. J Food Prot 74(5):718–726

Shen Q, Kong F, Wang Q (2006) Effect of modified atmosphere packaging on the browning and lignification of bamboo shoots. J Food Eng 77:348–354

Simon A, Gonzalez-Fantos E (2011) Influence of modified atmosphere packaging and storage temperature on the sensory and microbiological quality of fresh peeled white asparagus. Food Control 22:369–374

Simon A, Gonzalez-Fandos E, Tobar V (2004) Influence of washing and packaging on the sensory and microbiological quality of fresh peeled white asparagus. J Food Sci 69(1):6–12

Simon A, Gonzalez-Fandos E, Tobar V (2005) The sensory and microbiological quality of fresh sliced mushroom (*Agaricus bisporus L.*) packaged in modified atmospheres. Int J Food Sci Technol 40:943–952

Simon A, Gonzales-Fandos E, Rodriguez D (2008) Effect of film and temperature on the sensory, microbiological and nutritional quality of minimally processed cauliflower. Int J Food Sci Technol 43:1628–1636

Simon E, Gonzalez-Fandos E, Vazquez M (2010) Effect of washing with citric acid and packaging in modified atmosphere on the sensory and microbiological quality of sliced mushrooms (*Agaricus bisporus L.*). Food Control 21:851–856

Siomos AS, Sfakiotakis EM, Dogras CC (2000) Modified atmosphere packaging of white asparagus spears: composition, color and textural quality responses to temperature and light. Sci Hortic 84:1–13

Siomos SA, Gerasopoulos D, Tsouvaltzis P, Koukounaras A (2010) Effects of heat treatment on atmospheric composition and color of peeled white asparagus in modified atmosphere packaging. Innovative Food Sci Emerg Technol 11:118–122

Siripatrawan U, Assatarakul K (2009) Methyl jasmonate coupled with modified atmosphere packaging to extend shelf life of tomato (*Lycopersicon esculentum* Mill.) during cold storage. Int J Food Sci Technol 44:1065–1071

Sothornvit R, Kiatchanapaibul P (2009) Quality and shelf-life of washed fresh-cut asparagus in modified atmosphere packaging. LWT Food Sci Technol 42:1484–1490

Suparlan, Itoh K (2003) Combined effects of hot water treatment (HWT) and modified atmosphere packaging (MAP) on quality of tomatoes. Packag Technol Sci 16:171–178

Tano K, Oule MK, Doyon G, Lencki RW, Arul J (2007) Comparative evaluation of the effect of storage temperature fluctuation on modified atmosphere packages of selected fruit and vegetables. Postharvest Biol Technol 46:212–221

Tao F, Zhang M, Hangqing Y, Jincai S (2006) Effects of different storage conditions on chemical and physical properties of white mushrooms after vacuum cooling. J Food Eng 77:545–549

Tao F, Zhang M, Yu H (2007) Effect of vacuum cooling on physiological changes in the antioxidant system of mushroom under different storage conditions. J Food Eng 79:1302–1309

Tassou CC, Boziaris JS (2002) Survival of *Salmonella enteritidis* and changes in pH and organic acids in grated carrots inoculated or not with *Lactobacillus sp.* and stored under different atmospheres at 4°C. J Sci Food Agric 82:1122–1127

Tenorio MD, Villanueva MJ, Sagardoy M (2004) Changes in carotenoids and chlorophylls in fresh green asparagus (*Asparagus officinalis L.*) stored under modified atmosphere packaging. J Sci Food Agric 84:357–365

Toivonen PMA, Stan S (2004) The effect of washing on physicochemical changes in packaged, sliced green peppers. Int J Food Sci Technol 39:43–51

Valero A, Carrasco E, Perez-Rodriguez F, Garcia-Gimeno RM, Blanco C, Zurera G (2006) Monitoring the sensorial and microbiological quality of pasteurized white asparagus at different storage temperatures. J Sci Food Agric 86:1281–1288

Villaescusa R, Gil MI (2003) Quality improvement of *Pleurotus* mushrooms by modified atmosphere packaging and moisture absorbers. Postharvest Biol Technol 28:169–179

Villanueva MJ, Tenorio MD, Sagardoy M, Redondo A, Saco MD (2005) Physical, chemical, histological and microbiological changes in fresh green asparagus (*Asparagus officinalis L.*) stored in modified atmosphere packaging. Food Chem 91:609–619

Wall MM, Berghage RD (1996) Prolonging the shelf-life of fresh green chile peppers through modified atmosphere packaging and low temperature storage. J Food Qual 19:467–477

Wang CY, Qi L (1997) Modified atmosphere packaging alleviates chilling injury in cucumbers. Postharvest Biol Technol 10:195–200

Wang TC, Wang TC, Cao PY, Nout MJR, Sun GB, Liu L (2011) Effect of modified atmosphere packaging (MAP) with low and superatmospheric oxygen on the quality and antioxidant enzyme system of golden needle mushrooms (*Flammulina velutipes*) during postharvest storage. Eur Food Res Technol 232:851–860

Workneh ST, Osthoff G (2010) A review on integrated agro-technology of vegetables. Afr J Biotechnol 9(54):9307–9327

Workneh ST, Osthoff G, Steyn SM (2011) Influence of preharvest and postharvest treatments on stored tomato quality. Afr J Agric Res 6(12):2725–2736

Xing Y, Li X, Xu Q, Jiang Y, Yun J, Li W (2010) Effects of chitosan-based coating and modified atmosphere packaging (MAP) on browning and shelf life of fresh-cut lotus root (*Nelumbo nucifera* Gaerth). Innovative Food Sci Emerg Technol 11:684–689

Zagory D (1999) Effects of post-processing handling and packaging on microbial populations. Postharvest Biol Technol 15:313–321

Zhang M, Zhan ZG, Wang SJ, Tang JM (2008) Extending the shelf-life of asparagus spears with a compressed mix of argon and xenon gases. LWT 41:686–691

Zhuang H, Hildebrand DF, Barth MM (1995) Senescence of broccoli buds is related to changes in lipid peroxidation. J Agric Food Chem 43(10):2585–2591